SpringerBriefs in Service Science

SpringerBriefs present concise summaries of cutting-edge research and practical applications across a wide spectrum of fields. Featuring compact volumes of 50 to 125 pages, the series covers a range of content from professional to academic.

Typical publications can be:

A timely report of state-of-the art methods

A bridge between new research results, as published in journal articles

A snapshot of a hot or emerging topic

An in-depth case study

A presentation of core concepts that students must understand in order to make independent contributions

SpringerBriefs are characterized by fast, global electronic dissemination, standard publishing contracts, standardized manuscript preparation and formatting guidelines, and expedited production schedules.

The rapidly growing fields of Big Data, AI and Machine Learning, together with emerging analytic theories and technologies, have allowed us to gain comprehensive insights into both social and transactional interactions in service value co-creation processes. The series SpringerBriefs in Service Science is devoted to publications that offer new perspectives on service research by showcasing service transformations across various sectors of the digital economy. The research findings presented will help service organizations address their service challenges in tomorrow's service-oriented economy.

Youakim Badr

Smart Digital Service Ecosystems

A Research Roadmap from Service
Computing and Engineering Perspectives

 Springer

Youakim Badr
School of Graduate Professional Studies
Pennsylvania State University
Malvern, PA, USA

ISSN 2731-3743   ISSN 2731-3751   (electronic)
SpringerBriefs in Service Science
ISBN 978-3-031-27925-6   ISBN 978-3-031-27926-3   (eBook)
https://doi.org/10.1007/978-3-031-27926-3

This Springer imprint is published by the registered company Springer Nature Switzerland AG
The registered company address is: Gewerbestrasse 11, 6330 Cham, Switzerland

To Chloé, Raphaël, and Carmen

Preface

Setting out to write a monograph is not a conventional endeavor. Indeed, there is no manual, norm, written, or unwritten rules.

I started by asking myself one question: "For whom does one write a monograph?" To this question, two answers can be given, each of which guides me to organize my thoughts and structure the content.

– For researchers: Writing a hundred or more pages that will only be read by a small handful of readers is not necessarily effective. It is my hope that this monograph proves helpful to young researchers, starting out in discovering *Service Science*. This body of work also aims to provide experienced researchers with research directions and holistic approach from two different yet complementary disciplines: *service computing* and *service engineering* design and build IT-enabled services.
– For oneself: A monograph is also written and motivated by my desire to put into black and white a wealth of scientific contributions, and a research strategy built around funded projects and international research collaborations. Writing is the best way to crystallize ideas, increase memory and retention, and help us learn.

The reader may wonder if the presence of a section is justified or not, if the monograph's length is not too long, or if the tone is appropriate. In as much as possible, I have sought to write the chapters that can be read at several levels: from a quick reading while retaining titles and important points in bold, to a more thorough reading, for which numerous tables and figures illustrate and make the points clearer.

Malvern, PA, USA   Youakim Badr
November 2022

Acknowledgments

The body of work presented in this book could not have come to fruition without the support of Professor Robin Qiu. His seminal work on computational thinking of service systems intrigued my curiosity and introduced me to the Service Science discipline in 2010. Ever since this date, most of my research contributions revolve around service computing and service engineering. I also truly appreciate him taking time to read and re-read my work and the discussions, which followed.

I also would like to thank all those who, from near or far, helped me to thrive in my career and life. I would like to express my sincere thanks to colleagues at INSA-Lyon and The Pennsylvania State University, doctoral and graduate students, some of whom have become close friends.

I am also indebted to my children: *Chloé*, *Raphaël* and *Carmen*, for their unconditional love and unwavering support throughout my work. Their presence by my side brought me joy and serenity I needed to thrive and be the best version of myself. Their energy and joyfulness inspire me more than they will ever know.

Contents

1 The Rise of IT-Enabled Services 1
  1.1 The Service Economy... 1
  1.2 Productivity in Services.. 2
      1.2.1 Multivarious Definitions...................................... 2
      1.2.2 Multidisciplinarity of Services 3
      1.2.3 Unconventional Characteristics of Services 3
  1.3 Services in Service Science and the Service-Dominant Logic........ 4
  1.4 IT-Enabled Services.. 5
  1.5 Designing and Implementing IT-Enabled Services................... 7
      1.5.1 The Systemic Thinking Perspective 7
      1.5.2 The ICT Perspective 8
      1.5.3 Service System Challenges.................................... 9
          1.5.3.1 Compositions in Software Engineering:
                  The SOA Perspective............................ 11
          1.5.3.2 Compositions in Service Engineering: The
                  Business Perspective 13
      1.5.4 Challenges Related to the Service Concept.................... 15
      1.5.5 Challenges Related to Service Processes 17
      1.5.6 A Research Road Map for IT-Enabled Services............... 19

2 Service Reference Model and Requirements 21
  2.1 Introduction 21
  2.2 Services and Challenges 22
      2.2.1 The Service Concept.. 22
      2.2.2 Modeling Services 24
      2.2.3 Service Frameworks 25
  2.3 Design Requirements for Services 26
  2.4 An Architectural Framework for IT-Enabled Services 28
      2.4.1 Service System Reference Model............................. 29
          2.4.1.1 The System View 30
          2.4.1.2 The Characteristics View............................. 32

           2.4.1.3   The Business View 34

           2.4.1.4   The Interaction View 37

      2.4.2   The Service Shared Requirement Model 38

           2.4.2.1   Specifying Requirements with Service
Characteristics 41

           2.4.2.2   The Goal-Oriented Graphical Model 42

           2.4.2.3   The Service Requirement Modeling Language 42

           2.4.2.4   Customer Requirements with SBVR-Based
Templates 43

           2.4.2.5   Requirement Mapping Algorithms 46

  2.5   Concluding Remarks.. 48

3 Collaborative Design Methods Driven by Business Artifacts 49

  3.1   Introduction 49

  3.2   Challenges Related to Service Processes 50

  3.3   Service Collaborative Design and Processes with Business
Artifacts 53

      3.3.1   The Collaboration Model...................................... 54

           3.3.1.1   Interaction Patterns 56

           3.3.1.2   The Service Process Model 57

           3.3.1.3   The Collaboration Process Life Cycle.............. 60

           3.3.1.4   Collaboration Patterns............................. 61

           3.3.1.5   The Collaborative Service Design Method.......... 62

      3.3.2   The Business Artifact Discovery Method 65

           3.3.2.1   The Data Perspective 66

           3.3.2.2   The Operation Perspective 68

           3.3.2.3   The Common Perspective 69

  3.4   Concluding Remarks.. 70

4 Toward Digital Service Ecosystems 73

  4.1   Introduction 73

  4.2   Challenges Related to the Digital Ecosystems........................ 75

  4.3   The Digital Service Ecosystem 79

      4.3.1   Social-Based Relationships Between Service
Components 79

      4.3.2   The Evolution of Digital Service Ecosystems 81

      4.3.3   The Service Bundling Based on Collaboration Processes 82

           4.3.3.1   The SaaS-Based Architecture for Service
Front-Office Interfaces 86

           4.3.3.2   Managing Service Characteristics 88

  4.4   The Ad Hoc Web Service Composition............................... 88

      4.4.1   The Rule-Driven Composition Model 90

      4.4.2   The Service Farming Algorithm 92

  4.5   From Service Systems to Digital Service Ecosystems 96

  4.6   Web Service Selection 97

4.6.1   Web Service Selection Based on Nonfunctional
        Properties..   98
        4.6.1.1   Web Service Nonfunctional Properties..............   98
        4.6.1.2   WS-Policy Specification to Model
                  Nonfunctional Properties............................ 100
        4.6.1.3   Publishing NFP WS-Policies in UDDI
                  Registries 102
        4.6.1.4   The Matching Algorithm............................. 103
        4.6.1.5   Including User Preferences in Web Service
                  Selection 105

5   Services in the Era of Artificial Intelligence and Internet of Things ... 107
   5.1   Toward AI- and IoT-Enabled Services 107
   5.2   Internet of Things and Services Systems 108
         5.2.1   Self-Adaptable Connected Devices 109
         5.2.2   IoT Big Data Management and Built-in Analytics............ 110
   5.3   Artificial Intelligence and Services Systems......................... 112
         5.3.1   Composable AI Service Systems and Security-by-Design ... 113
         5.3.2   AI Risks in Service Systems 114
         5.3.3   Privacy-Preserving and Resilient Federated Learning 115

References.. 117

Chapter 1
The Rise of IT-Enabled Services

Keywords Tertiary sector · Sector of services · Service science ·
Service-dominant logic · Service characteristics · IT-enabled services · Service
concept · Service processes · Service systems · Service-oriented architecture ·
Web services · Service engineering · Systemic thinking · Systems theory · Web
service composition · Web service S · Business services · Service bundling ·
Service characteristics

1.1 The Service Economy

The tertiary sector, known also as the sector of services, drains most of the economic
workforce and produces the highest part of gross national products (GNP) in
developed countries; the 2017 estimation [1], for example, shows that services
comprise nearly 80% of the US's GDP and 78.8% of the France's GDP. Compared
to the manufacturing and agricultural sectors, the service sector consists of soft parts
of the economy, that is to say, activities where individuals offer their knowledge and
time to improve productivity, performance, potential, and sustainability. Services
as intangible goods include attention, advice, experience, and solutions. A wide
diversity of services exists today, including healthcare, education, finance, transport,
communication, information technology, tourism, R&D, engineering, government,
security, insurance, retail, leisure, utilities, entertainment, transportation, and professional services, just to mention a few.

The service sector has grown the most globally in recent years and has emerged
as the main source of job creation, often compensating for job losses in manufacturing [2]. Several reasons explain the trend of its development [3]:

- The increase in personal income, whereby individuals increasingly are willing to
  pay for their chores to be done by somebody else.
- The growth of aging population, which led to the emergence of new assistance
  businesses.

© The Author(s), under exclusive license to Springer Nature Switzerland AG 2023
Y. Badr, *Smart Digital Service Ecosystems*, SpringerBriefs in Service Science,
https://doi.org/10.1007/978-3-031-27926-3_1

- The technological development, such as advances in Internet and artificial intelligence technologies, has dramatically improved the quality of available services at large scale.
- The introduction of advanced products, which led to the increase in demands for support and maintenance.
- Outsourcing of business processes, which led companies to concentrate on their core competence and businesses and relocate to third parties all activities that are low-added value and irrelevant to their core businesses.

From a broader perspective, some experts debate that the reason for the rise of the service economy focuses on the shift from ownership and trade of land and natural resources to one that is driven by knowledge, skills, and other human assets. This shift is now widely known as the beginning of the Information Age (or Knowledge Age) to distinguish it from the Industrial Age [4]. Paradoxically, compared to the manufacturing sector, many studies affirm that service innovation is struggling, and service productivity is rather low, since service offerings do not meet the growing demands despite the constant call for services' improvement in most advanced countries [5].

1.2   Productivity in Services

The main factors for the low productivity in services include multivarious definitions, multidisciplinarity of services, and unconventional characteristics of services. In the following sections, we discuss the challenges that lead to low productivity in services.

1.2.1   Multivarious Definitions

Despite many efforts, the definition of "*what is a service?*" is still an intensive debate [6]. An interesting listing of most used definitions in the service literature is reported by Fitzsimmons and Fitzsimmons [7], which shows that some scholars have decided to define services by establishing taxonomies and grouping them into several categories whereas others consider services as the antithesis of goods and attempt to compare them. There are numerous definitions ranging from "*everything that cannot be dropped on your feet*" to describing services as processes, acts, deed, and performances [8]. Since it is almost impossible to have a consensus in the service literature, we observe that most definitions share a common concept that is "*pay for performance in which client and provider coproduce value.*" The performance can range from high talent performance to high technological performance. Nevertheless, we observe that the notion of coproduction of value is implicitly or explicitly present in most definitions. The coproduction results from the collaboration between service providers and service consumers to jointly design and consume services. The value refers to mutual benefits for providers and consumers. Opitz and Schwengels [9] put the notion of *"coproduction"* even more pithily and apply the terms coinventor, codesigner, cocreator, and coproducer to refer to the active involvement of customers. The coproduction reveals a particular challenge in which services are rightfully people-centric, and they should be jointly developed with people and for the sake of their satisfaction.

1.2.2 *Multidisciplinarity of Services*

The proliferation of services in almost all business domains leads to the lack of a unified view of the "service" concept, which is extensively used with different meanings and perspectives in at least six distinct disciplines: service computing (computer science) [10], service engineering (service science) [11], service marketing [7], service operations [12], and service management [3]. The disagreements and conflicting views of services between and within these disciplines show obstacles to develop and innovate in services and, consequently, cannot facilitate effective communication between service stakeholders. A crucial aspect of services is their socio-technical orientation, which combines people, technologies, and business activities, resulting into complex socio-technical systems [13]. Moreover, the multidisciplinary nature of services becomes a barrier to their development across and within different disciplines [14], p. 10.

1.2.3 *Unconventional Characteristics of Services*

Services have unconventional characteristics that distinguish them from goods. These characteristics commonly known as IHIP (i.e., intangibility, heterogeneity, inseparability, and perishability) and have profound impact on service innovation and productivity [15].

Intangibility: refers to the state whereby services are inaccessible to senses and lack the palpable or tactile quality of goods. Services provided as a result of the provider's activities are being less concrete than of tangible products. Therefore, it is more difficult to price and manage services, define and measure productivity, and evaluate service quality. Lovelock and Gummesson [16] argue that mental or physical intangibility results in difficulties for customers to evaluate services before purchase.

Heterogeneity: refers to the state whereby the same service can have different effects and produce different responses depending on their provider, place of provision, and user's mental state and environment. Being heterogeneous implies that each customer subjectively evaluates the service outcome. Heterogeneity leads to diverse evaluations for the same service from different customers and requires service quality improvement.

Inseparability of production and consumption: In goods, production, distribution, and purchase activities are separated from good usage in both time and space. While services are simultaneously produced and consumed, production and consumption are interactive and inseparable processes, which require the simultaneous involvement of providers and customers to design and consume services. The customer becomes a central part of the service production and consumption, and consequently, the inseparability arises problems of how to improve the customer's capacity to be effectively involved in services.

Perishability: means that services cannot be saved, stored for reuse later, resold, or returned as compared to goods. This implies that services cannot be inventoried, and unused services are thus considered wasted. The perishability is associated to the intangibility characteristic. Goods not sold have no value and services without consumers cannot be produced. The value cocreation is only possible when services are consumed by customers [16].

Based on the abovementioned three factors, the service landscape is unclear and arises global challenges on "how to understand, design, deliver and improve services" to increase productivity and innovation in the service sector. While we are aware of the complexity of these challenges we face, the literature on services often shows a lack of theoretical backgrounds in services [14].

In the last decade, a new mindset in understanding and designing services has emerged, thanks to some remarkable efforts, notably the service-dominant logic paradigm and the service science initiative, as well as the latest advances in information and communication technologies (ICT). All together thereby give the birth to "IT-enabled services" that are information-driven, customer-centric, e-oriented, and productivity-focused services enabled by ICT [17].

1.3   Services in Service Science and the Service-Dominant Logic

The service-dominant logic (S-DL) paradigm, as introduced by Vargo and Lush, claims that service (in the singular), as the core concept, replaces both goods and services, and they assert that *"all economies are service economies and all businesses are service businesses. Service is the application of competences for the benefit of another entity, a service is exchanged for a service, value is always co-created and goods are only appliances for service delivery"* [18]. In this logic, service providers offer a value proposition, and the value becomes the outcome of the cocreation activity between providers and customers. The customer is always a coproducer and provides significant inputs into service design and consumption processes [19]. This new mindset, as summarized by Bill Hefley et al. [20], has a direct impact on our practice, as computer scientists, in developing solutions for businesses and, particularly, in using information and communication technologies (ICT) to develop software applications and infrastructures to support real-world services. This new line of thinking implies that the service-centric logic should guide the design and implementation of new service systems focused on people, driven by information sharing, created through collaboration, and enabled by ICTs. The S-DL had an impact on computer science with the proliferation of approaches, calling for everything as a service (i.e., SaaS, Internet of services) [21].

The service science (SS) initiative, short for *services sciences, management, and engineering* (SSME), was formally introduced by the CEO of IBM, Samuel Palmisano, in December 2004 to promote an interdisciplinary research in services in collaboration with universities, companies, governments, and research organizations [22]. Service science advocates an interdisciplinary study of computer science, mathematics, the theory of decision-support, social and cognitive sciences, and other fields to address the shortcomings of complexity, productivity, innovation, and industrialization of services [23]. Service science also promulgates an academic discipline to study services systems. The term "service system" is the fundamental concept in service science, whereby specific arrangements of skills and technologies take actions that produce added-value solutions for business problems. The service system is often defined as "*... a configuration of people, technologies, and other resources that interact with other service systems to create mutual value*" [24]. What should be researched in SS is still under debate and thus remains largely undefined. More and more research centers worldwide and prestigious universities (i.e., UC Berkeley, Harvard University, MIT) are devoted to establish the foundations of service science; the majority of which are located in the USA and UK [6].

A growing body of literature in services appeals theoretical insights from S-DL and service science to gain an understanding of the factors that can contribute to design services [25, 26]. Both S-DL and SS contend that the integration of skills, needs, resources, information, and business goals among providers and users incite service systems [27]. Despite many solid contributions and recent initiatives (e.g., SRII, Cambridge SSME Symposium [14]), the service research field remains highly fragmented, leading to disagreement in vocabularies, methods, tools, and models. The design of services remains balkanized without an integrative approach to build a multidisciplinary ecosystem for service innovation, yet recent ICT advances open new opportunities to design and develop services [23].

1.4 IT-Enabled Services

As the world becomes more and more connected economically, technically, and socially through advanced ICTs, many of today's services tend to aggregate customized and intelligent solutions through the development of integrated software and hardware devices [28, 29]. "*[...] almost anything from people, object, to process, for any organization, large or small – can become digitally aware and networked* [30]". The integration of ICTs in designing and developing service systems and their delivery channels have led Dolly Bhasin to define "IT-enabled services" [31]. In her words, "when information technology enables a (real world) service by improving its quality, increasing customer satisfaction and adding value, it is an IT-enabled service." IT-enabled services cover an entire gamut of new services, which include opportunities in smart cities, smart home, living labs, Internet of services, e-learning, App Stores, home healthcare, knowledge-intensive firms, and open innovation.

IT-enabled services share common characteristics as they are designed and developed in ICT-enabled environments, which we define them as digital, social-based, collaborative, ad hoc, dynamic, distributed, and opened environments. These characteristics encourage introspection of behavior and communication, but they also imply new challenges such as new computing models and capabilities at a scale never before imagined (i.e., digital ecosystems). As a matter of fact, information technologies, especially Web 2.0, lead to the emergence of social networks and virtual communities made up of individuals and organizations that actively interact and collaborate with each other as creators of information and online services in contrast to Web 1.0, where people are limited to reading information from websites [32]. Collaboration has facilitated the development of complex solutions in ICT-enabled environments, integrating social and economic aspects, and information technologies. Casual users can become service stakeholders in such digital environments and play active roles by producing almost everything as services (*-as-Services) [33]. The shift of delivering services through the Internet postulates novel digital ecosystems with promises of sustainable business development and new business opportunities. IT-enabled services thus inspire new research visions to increase service productivity and innovations [34, 35]. They also enable small- and medium-sized enterprises with new business models and markets [36].

Tien and Berg have initiated in 2003 the study of IT-enabled services (a.k.a emergent services) and their design from the system engineering perspective. They assert that " [...] service systems engineering methods should address the information-driven, customer-centric, e-oriented and productivity-focused issues that are pertinent to services [17], p. 33". In this perspective, a possible design method for IT-enabled service systems is thus driven by a coproduction experience (e.g., options, controls, decisions, ...) of both providers and consumers, and it results in no separation between service production and service delivery. It is worth noting here that IT-enabled services should not be confused with IT services and technological services. According to Chau et al. [37], IT services support the digital transformation of traditional business activities into IT-software applications to automate business processes, create delivery channels (e.g., via Internet and smart phones), and/or facilitate collaboration among various actors. This vision makes IT services (produced as products originated in software engineering to support business activities) and services enabled by IT (produced as interactions originated in the fields of S-DL and SS in order to cocreate services) two different paradigms and are not interchangeable in the sense that services enabled by IT are not IT services.

IT-enabled services have received an abundant attention only from a techno-logical perspectiveas they are often reduced to suitable e-services [38] or Web services [39] (i.e., technological services). This is a narrow context in which e-services or Web services only focus on user-software interactions to process information and automate tasks. It turns out that IT-enabled services cover a wider scope, exceeding technological services so far by focusing on business aspects and value cocreation. While technological services can be realized with service-oriented architectures and Web services, IT-enabled services can be achieved by collaboration among service stakeholders to satisfy customer needs, improve service quality, and increase customer satisfaction in social-centered, ad hoc, and distributed environments.

1.5 Designing and Implementing IT-Enabled Services

If all businesses can be seen as service businesses under the premise of the S-DL paradigm and if those businesses are produced by systems as postulated by the SS, a deep understanding of the service system might lead to deeper and fundamental understandings of service design and innovation. Kim and Nam [40] argue that "*only through the lens of systems, researchers in service literature can recognize the big picture of service and its components and find an appropriate way of improving and innovating it*". This leads us to investigate service systems from two perspectives:

1.5.1 The Systemic Thinking Perspective

Skyttner [41] clarifies that "*A system is a set of interacting units or elements that form an integrated whole intended to perform some function*". The system is thus perceived as assemblage of interrelated elements and comprises a unified whole to enable the flow of resources to achieve a specific goal. All systems have common characteristics such as:

1. System structure, which is defined by elements and their composition
2. System behavior, which comprises how inputs are processed into outputs
3. System interconnectivity, which is defined by relationships among various system elements
4. System functions, which are achieved by the operations that the system performs
5. Cybernetics characteristic, which includes feedback loops (i.e., evaluation of performance) and control loops (i.e., self-regulation, adaptation, optimization, and/or management) to ensure the system sustainability [42]

The systems theory makes it possible to apply two ways of thinking to gain understanding on service systems: Firstly, the systems thinking theory, which is about perceiving or understanding how things, regarded as systems, interacts with (affects and is affected by) the things around them. Secondly, the systemic thinking approach (i.e., thinking in terms of wholes) seeks to gain insights into complex situations and problems by focusing on the system elements that will improve the entire whole system the most, instead of elements that can be improved the most [43].

1.5.2   The ICT Perspective

Wing and Qiu agree that applying a computational thinking to service systems enable problem-solving and designing systems at a scale never before imagined [28, 44]. Computer science is an appealing "science" that manifests abstraction abilities and computational capabilities by applying models to represent anything, process and exchange information, and work efficiently and collaborate with anyone, anytime, and anywhere. In fact, software are ultimately intangible products representing ideas, artifacts, procedures. eComputer science thus plays a pivotal role by advocating an interdisciplinary approach among service disciplines and creating a body of knowledge to bridge or integrate all service disciplines (e.g., service operations, service management, etc.).

Through this book, we are interested in challenges and solutions related to the design and implementation of IT-enabled service systems from the systemic thinking perspective and within the framework of the service-dominant logic paradigm and the service science initiative. Before discussing more precisely the challenges that we have confronted in our research activities and projects and present later our contributions to overcome these challenges, we firstly have to delineate our understanding on the "service design process" in general to better understand what we mean by the design of IT-enabled services.

The service design, at its heart, is about the "*specification of an object, manifested by an agent, intended to accomplish goals, in a particular environment, using a set of primitive components, satisfying a set of requirements, subject to constraints* [46]." This definition refers indirectly to the systemic characteristics (e.g., agents, goals, requirements, components, …) and has strongly motivated us to analyze what are the prerequisites for a potential design method for IT-enabled services and how to support the implementation of their systems from the ICT perspective. Edvardsson and Olsson identify three prerequisites to design services [47];

1. The service concept, which consists of what is to be done for the customer
2. The service process, which consists of the activities that are needed to create the service
3. The service system, which refers to resources that are required in service processes to realize benefits for customers

The precursor work of Edvardsson and Olsson on the service design model (see Fig. 1.1) aims to build foundations for the service design process. As explained

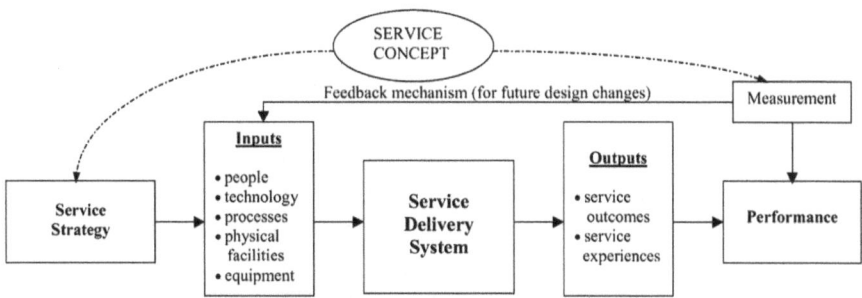

Fig. 1.1 A service design model (adopted from [45], p. 126])

in the following sections, most works in the service literature gravitate in general around services, service systems, and processes without, unfortunately, a holistic view or clear links between the service concept, service processes, and service systems.

Table 1.1 identifies service concept, service processes, and service systems as prerequisites for designing and implementing IT-enabled services and classifies research and implementation challenges from two disciplines: software engineering (service-oriented architecture perspective) and service engineering (business perspective systemic perspective).

1.5.3 Service System Challenges

According to Edvardsson and Olsson [47], the service system is a pillar prerequisite in service design methods. While the "service system" term frequently appears in the service literature [48–54], it is rarely well-defined and somewhat misleading by its lack of precisions. Despite apparent significant efforts, a common understanding on the service system structure and how it is possible to identify and assemble system elements is not yet available. According to Maglio et al. [22], a service system is "... *an organizational and socio-technological system embedded in complex and adaptive combination of resources and stakeholders: intermediaries, competitors, friends, government, and so on.*" This implies that the genesis of an IT-enabled service system is an "adaptive combination" of resources, consisting of two aspects:

1. The integration of system resources (e.g., tangible/intangible resources, business strategies, knowledge, competencies, ...), technologies (e.g., processes, software, devices, ...), and people (e.g., customers, developers, designers, deciders...) to create ICT-enabled environments and foster collaboration

Table 1.1 Challenges of designing and implementing IT-enabled services

	Research and implementation challenges			
Service concept	Why a service modeling framework is not available yet for service systems?			
Service processes	Why business processes are necessary but not sufficient to build service processes?			
Service systems	Software engineering (SOA perspective)		Service engineering (business perspective)	
Web service composition	How to compose Web services in ad hoc and dynamic environments?		Framework	Why extended SOA-based frameworks fail to deliver IT-enabled services?
Web Service selection and discovery	How to enhance Web service selection and discovery in social-based environments?		Bundling services	Why bundling services is insufficient to IT-enabled services?
Artificial intelligence	How to embed intelligence into service components? How to build Intelligent services?			

2. The service adaptability to respond to changes (e.g., new requirements or decisions) to decide on best strategies/actions to satisfy needs, solve problems, and cocreate value at each step during collaboration

Achieving "adaptive combinations" thus requires suitable "composition mechanisms" to integrate resources, technologies, and people to create service processes. Ensuring adaptability also requires the IT-enabled service system "feedback loops" (i.e., cybernetic characteristic) to enable service processes to sense changes or variations. A "composition mechanism" should thus recombine resources, technologies, and people ingenuity to attenuate variations by maintaining an "equilibrium" between customer needs and provider offerings (i.e., adjust the collaboration strategy to ensure that service is conceived as the customer perceives it) or by achieving new goals (i.e., new collaboration strategy to increase mutual benefits). Keep in mind the service IHIP characteristics; the main source of complexity in designing IT-enabled services is the difficulty of developing an assembly mechanism to build patterns across resources, people, and technologies [43] that improve the service system as a whole to effectively solve customer problems, exceed customer satisfaction, and create business benefits (cf. systemic thinking).

The service system concept leads us to the challenge of building the IT-enabled service system "as a combination" of resources and ensuring IT-enabled service adaptability (i.e., dynamicity) through feedback loops. The "logic of combinations" already applied in the disciplines, such as electronics and software engineering (a.k.a composition mechanism) and business services (a.k.a bundling mechanism). The composition mechanism aims at quickly building software applications from existing software packages rather than creating them from scratch, whereas the bundling mechanism aims at including two or more separate services in one package, often for a price that is lower than the sum of the prices of the separate services in the package. In our research activities, we focus on the logic of composition in software engineering at the technical level and business services at the business level to design and implement IT-enabled service systems with reusable resources through collaboration in ICT-enabled environments. As demonstrated in the following sections, the logic of composition reveals drawbacks and limitations when we confront them to IT-enabled services in ad hoc, social-based dynamic and open environments.

1.5.3.1 Compositions in Software Engineering: The SOA Perspective

Over the last four decades, the practice of software development has gone through different software architectures (i.e., client/server, three-tier, n-tier, etc.), each of which was made in part to deal with greater levels of software complexity. Technologies, such as EJB, NET, and CORBA, are effective ways of implementing software components and allow developers to create more complex, higher-quality software systems faster than ever before. The service-oriented architecture (SOA) is an architectural style that has special design principles, such as loosely-coupled, contract-based, asynchronous communication, remote execution, etc. It is made up of software components and interconnections that stress location transparency and interoperability [55]. The SOA builds agile software applications from independently software components—called services. The term "service" is misused here. From the computer science perspective, a (technological) service is component-based software, self-contained, and self-described with an interface that controls access to its functionalities. Technological services, as basic constructs, offer solutions for enterprises to build agile and interoperable information systems, spanning organizational boundaries.

The SOA is not new and has been used for years. However, what is relatively new is the emergence of Web services [56]. The W3C consortium defines the Web service as "*a software system designed to support interoperable machine-to-machine interaction over a network.*" A Web service may be fully described with functional and nonfunctional properties. Functional properties define what the service *can do,* whereas the nonfunctional properties depict how the service can using service operations. With Web services and their *de facto* standards (e.g., WSDL, SOAP, and UDDI), there has been a renewed interest in the SOA, since the predominant implementation of SOAs can be found today in Web services. Together, Web services and SOA are ultimately about designing and building business processes and applications by coordinating the execution of multiple Web services [57]. Web service composition is the process of reusing existing Web services and logically recombining them into composite services. A composite Web service provides new functionalities that existing Web services cannot provide alone. By such, SOA guarantees reusability [58], interoperability [59], scalability [60], and cost efficiency [61], which are essential characteristics of modern enterprise information systems.

Constructing composite Web Services while meeting user requirements on both functional and nonfunctional properties requires three consecutive phases [62]:

1. **Web service discovery phase** corresponds to the activity of locating Web services that meet requirements on their functional properties. Based on WSDL and UDDI registries, many algorithms, tools, and mechanisms are proposed to enhance the accuracy of Web service discovery by applying syntactic matching (keyword-based matches [63]) and/or semantic matching (i.e., ontology-based discovery [64], machine learning [65], information retrieval [66], semantic annotation [67], etc.).
2. **Web service selection phase** refers to the activity of evaluating and ranking the discovered Web services that fulfill requirements on their nonfunctional properties (i.e., quality of services properties). Different approaches have focused on QoS ontologies [68] and ad hoc methods within general Web service frameworks [69–73].
3. **Web service composition** refers to the interactions between Web services, which can be achieved with choreography and orchestration [74]. The choreography describes the external visible behavior of Web services from the service consumer point of view, whereas the orchestration describes how Web

services cooperate to achieve a common goal. The literature shows a plethora of languages specifying the exchange of messages between Web services prior to their execution, such as WSFL, XPDL, WSCI, WS-CDL, YAWL, and WS-BPEL [75, 76].

Without doubt, SOA and Web services provide noteworthy computational infrastructures to support IT-enabled services as well as the logic of composition. Unfortunately, IT-enabled service characteristics (i.e., information-driven, customer-centric, e-oriented, and productivity-focused [17]) and ICT-enabled environments (i.e., collaborative, social-based, ad hoc, dynamic, distributed, and opened environments), in which IT-enabled services are designed, developed, and consumed, arise the following new challenges to Web service composition, discovery, selection, and information security never handled before.

1.5.3.2 Compositions in Service Engineering: The Business Perspective

After discussing the composition mechanism and its challenges from the software engineering perspective with emphasis on Web service discovery, selection, composition, information security, and SOA design methods, we focus now on the logic of composition from the business perspective to assemble resources in IT-enabled service systems. In business management, the term "bundling" is often used to integrate two or more services into a new service, often for a price that is lower than the sum of the individual services [77]. We address two main questions when bundling services in ICT-enabled environments:

1. Could Web service composition mechanisms evolve from fine-grained software components (e.g., Web services) to compose large-grained components (i.e., SOA-based systems/subsystems) to create new service systems or bundles?
2. How bundling mechanisms in services can be applied to IT-enabled services?

The following subsections survey research fields in attempt to answer these questions; the former attempts to extend SOAs from a software perspective toward the service-oriented business perspective, whereas the latter has sought to build IT-enabled services as bundles of resources to satisfy customer needs and cocreate value.

Limitations of SOA-Based Frameworks

Most of research on SOAs takes a software engineering perspective and consider the design of business services or e-services as a software design problem. This statement is partially true and partially false. In fact, recent efforts struggle to leverage SOA design methods and/or rely on architectural frameworks to develop business services. Architectural frameworks, such as SOMA [78], CBM [79], SEASIDE [80], ISE [80], and MRM [81], are often inspired by the Zachman enterprise framework [82] and intend to provide a highly structured way of viewing and defining enterprise businesses. As an example of the transition from software engineering to the development of business services, we cite the extension of the service-oriented analysis and design method (SOAD) [83] to build the service-oriented modeling and architecture (SOMA). In fact, the goal service model (GSM) in the SOMA provides a bottom-up composition approach and helps to define the business service model and how to deliver it. Yet another example is the component business model (CBM) framework, which takes a top-down approach to create business services. The CBM makes a major shift by considering the entire enterprise and breaking it into separate basic building blocks (e.g., people, processes and technology) that can be combined together [84]. In our view, top-down approaches based on the Zachman framework and bottom-up approaches are quiet far from designing and developing IT-enabled services, and they show severe deficiencies. Firstly, business services are only assumed to be assemblages of components that match customer requirements. This assumption quickly reveals a drawback in practice since customer requirements are often unclear and incomplete, and they require negotiations and trade-offs. Secondly, the role of customers is only limited to providing requirements as inputs in the early design stage and exploiting later the produced services without any active involvement in the service design and development.

Bundling services aims at finding trade-offs between customers wants and providers' capabilities within a business context to maximize profits for both parties [77]. In IT-enabled services, service design and consumption imply collaboration, trade-offs, and negotiations and impose constraints on resources and assets related to service providers and consumers. Intangible resources (e.g., information, skills, processes, ...) or tangible resources (e.g., devices) are scarce and often cannot be shared and condemned to be in one place at a time (e.g., keeping know-how) [85]. Feedbacks from service stakeholders and adaptation to changes are also important aspects of IT-enabled services. In practice, what we observe in the SOA is that Web service composition mechanisms, which are used to assemble business activities accessed through Web services, primarily aim at reconciling incompatible (input and output) messages between Web services in a well-defined business context. Based on these arguments, describing business activities with only Web services and basic operations, such Web service selection, discovery, and composition, is convincingly insufficient operations to support service bundling or to create new IT-enabled services.

In other words, the SOA-based frameworks fail to deliver IT-enabled services. So far, what is lacking is a new shift from static Web service compositions in well-defined business contexts and unchanged constraints to dynamic compositions in ad hoc environments to deal with changes and variations (e.g., feedback, new needs), and recompose Web services accordingly. To support service bundling in collaborative environments, novel composition logic should capture customer participation and feedback to continuously improve service processes and customer satisfactions; Web service compositions should also be capable of considering multiple constraints on resources related to Web services and compose Web services.

Limitations of Service Bundling Approaches

Yet another important question should be whether bundling services would be sufficient to design and build IT-enabled services. Based on collaboration among service participants and exchange of resources, we found few works that tackle the problem of IT-enabled service bundling from the business perspective. These works mainly accentuate customer-centric or resource-centric design methods. For example, Pistore et al. [86] focus on resources as any object that is regarded as valuable and controlled by an agent, and on events as activities, affecting resources. They consider a service, as a resource, produced by one agent for another agent using certain capabilities (or other resources). They also develop a customer-based service bundling to mash-up widgets (i.e., Web services), which are considered merely interfaces to real-world services. Baida [77] takes a different approach and focuses on bundling services by means of aided-software tools to find trade-offs between customer demands (i.e., feature space) and resources (i.e., solution space of available services). He matches the service value perspective (e.g., needs, wants, and demands) with the service offering perspective (i.e., resources) to bundle services. Qiu [28] adopts a computational thinking approach to model dynamics and adaptiveness of IT-services service systems and capture customers' requirements, changes, expectation, and satisfaction as a variety of technical, social, and cultural aspects. Unlike SOA component-based compositions, these representative works show us that bundling IT-enabled services as combinations of resources reveals new challenges from the business perspective as how to manage scarce resources, trade-off demands and capabilities, select and bind IT-enabled services, adapt to changes, and improve customer satisfaction in ongoing collaboration.

From a business perspective, a novel bundling mechanism for IT-enabled services requires a holistic vision and foundations to easily integrate existing services into new services and scale them up to build a sustainable digital ecosystem of IT-enabled services. Such a mechanism also should not neglect all aspects inflicted by ICT-enabled environments, such as social-based Web service discovery, ad hoc composition in dynamic environments, and end-to-end secured SOA design method.

1.5.4 Challenges Related to the Service Concept

Edvardsson and Olsson [47] argue that the service concept—the second prerequisite in their service design model—refers to a prototype defined as the "*detailed description of what is to be done for the customer (i.e., the strategy to apply to satisfy customer needs and wishes) and how this is to be achieved (i.e., the configuration of resources)'.*" Service design methods have originally found in the service management discipline and organized into several categories, such as empathic design methods, participatory design methods, ethnography design methods, and codesign methods [87]. In the last decade, service design methods tend to focus on specific parts of services instead of the entire service experience (i.e., product-service bundles [88], human-centered services [89], knowledge-intensive services [90], user experience [91], ...). Most of these methods are qualitative, tactical, and operational based on best practices and observations. They often produce blueprints, scenarios, and use cases to visually describe sequences of interactions between service providers and consumers. Goldstein et al. [45] discuss the limitation of existing methods and design processes and argue that they adequately handle operational and tactical issues in service design, but they fail to refer to shared understanding of the nature of the service to be provided. They have defined the service concept by incorporating the necessary service elements as proposed by Chase and Bowen [92] (i.e., people, technology, processes, physical facilities, and equipment). Based on a shared "service concept," it becomes possible to move a step toward a systemic design by:

1. Defining a service strategy of what to deliver (e.g., market position and types of customer relationships) and how that strategy should be implemented to transform inputs to outputs by means of service delivery systems (i.e., service processes).
2. Setting up measurements on outputs to sense variations and feedback loops to adjust inputs. Ponsignon et al. [93] have studied the transformation of inputs to outputs (a.k.a the service delivery systems). They correlate input types (i.e., material, information, and customers) with transformation types to provide archetypical forms of service delivery systems.

Finally, a shared "service concept" by all service participants facilitates collaboration and helps addressing service design strategy and design decisions. It also enables systems thinking principles, by correlating service strategy to service performance (measurements and feedback loop) and by transforming service inputs to service output (service processes).

The service concept leads us to inspect in our research activities what are the basic elements (resources) in IT-enabled service systems and how they can be modeled and used to build service systems and apply a systemic thinking to find solution patterns. The identification of service system elements (resources) and their representational models are fundamental constructs in the development of any collaboration-based IT-enabled service design method that relies on the service strategy (the *what*) to build service delivery systems (the *how*) through Web service composition and bundling mechanisms. Since service delivery systems and processes will be discussed in the next section, we examine hereafter modeling frameworks, which intend to "concretize" the service concept and reduce its complexity by identifying necessary resources with their representational models.

From an IT-enabled service perspective, existing architectural frameworks do not fit well the purpose of describing, constructing, or implementing IT-enabled service systems for many reasons: Firstly, transformation strategies (e.g., information system urbanization [94]) in most architectural frameworks aim at only improving the information system performance by automating routine works, applying standards, and optimizing business processes. Secondly, current architectural frameworks do not provide vocabulary, models, and languages that help to identify service resources and their causal dependencies, which are important elements in the systemic thinking to define problem-solving patterns based on cause-effect relationships among service elements. Thirdly, service systems continuously require best collaboration strategies—instead of transformation strategies—to involve customers in service cocreation and consumption, satisfy their requirements, and improve the overall service quality. The shift from a transformation strategy to a collaboration strategy is justified by customer involvement in service processes to create services. Schekkerman concludes that "... *most enterprise architecture efforts are too inwardly focused, and do not include the customers and key business partners. This results in miss-aligned enterprise architectures, and lost opportunities to gain competitive advantage.*" Although the discipline of service science is still in its infancy, most architectural frameworks dedicated to service systems are mainly available in conference proceedings without a comprehensive list of service elements and without practical design methods to build IT-enabled services [27, 95–97].

Understanding, analyzing, and designing IT-enabled services require a shared service modeling framework, comprising common vocabularies to share domain knowledge, reference models to define set of views focusing on specific concerns, and modeling languages to provide constructs for reference models [97]. Such framework is mandatory to manage IT-enabled service systems from different point of views (business, information, collaboration, and infrastructure views), different perspectives (managers, customers, designers, IT professionals, developers, ...), and different levels (strategic and operational).

1.5.5   Challenges Related to Service Processes

After discussing challenges related to the service system and the service concept, we examine service processes—the third prerequisite in the service design model [47]. Roughly speaking, service processes, as sets of ad hoc activities, result from collaboration between service stakeholders. Unlike workflows in computer science or business processes in business management, which use predefined process control flows to determine what activities should be done, service processes focus on what can be done to achieve a business goal. Service stakeholders (i.e., knowledge workers) decide on how the goal is reached, and the role of service processes is thus assisting rather than guiding stakeholders in doing so [98]. Due to the information-intensive nature of service processes, they are ad hoc and incremental processes; they repetitively require inputs from service stakeholders to decide on appropriate strategies to perform [99]. Service processes also involve low level of repeatability, making them difficult to be automated by traditional process automation (e.g., business processes).

Congram and Epelmanargue that existing service design methods tend to produce flow diagrams to describe service processes [100]. Flow diagrams are used in different fields (i.e., service design, software engineering, and manufacturing) and cover a wide spectrum, ranging from blueprint diagrams, flowcharts, and control flow diagrams to functional flow block diagrams [101, 102]. Some diagrams work best at "big picture" thinking, while other focus on specific service details. Perhaps the Unified Services Theory (UST) [103] is the most interesting paradigm that focuses on service processes as the unit of analysis in service systems. The UST is formally stated as follows: "*With service processes, the customer provides significant inputs into the production process. With manufacturing processes, groups of customers may contribute ideas to the design of the product, but individual customers' only participation is to select and consume the output.*" By such, the presence of customer inputs is a necessary and sufficient condition to define the service process by which the customer actively takes part in the service design and delivery. The active participation of customers in service design and delivery is often regarded as the most salient difference between manufacturing processes and service processes. According to Shostack [104], modifications to service processes can increase or decrease process structural complexity and make the service more or less divergent. One problem is that, as yet, expressing and specifying processes in IT-enabled services with rigorous (technical) specifications as proposed by software engineering or as found in business processes can significantly damage customer participation and satisfaction. As a result, flow diagrams and business processes are not adaptable to high-variance customer inputs and cannot properly handle services processes [105].

Without excluding their utilities in automating structured and controlled routine activities, business processes show drawbacks and limitations when they are applied to service processes, for which there are many variations in customer inputs and ad hoc activities performed by service providers.

Business processes as activity-based processes are necessary to support service processes, but they are not sufficient to build effective service processes for the following reasons. Firstly, services are seen as information-intensive and people-intensive activities rather than routine business activities [25, 50, 90]. Tien and Berg assert that IT-enabled services are information-driven, customer-centric, e-oriented, and productivity-focused systems [17]. Service processes are thus information intensive processes, in which knowledgeable workers progressively and incrementally decide on actions to achieve common goals. Secondly, the role of ICT in IT-enabled services is rather assisting to create innovative solutions rather than automating business processes. Unlike business processes, which are driven by predefined control flows to determine "how" the work should be done, service processes focus on "what" can be done to achieve business goals, such as satisfying customer needs and improving the service quality.

There appears to be a severe gap between the promise of business processes of facilitating interactions between activities and what they really offer for services in terms of collaboration in ad hoc environments. The challenge of building service processes leads to investigate solutions for a truly paradigm shift from activity-oriented to information-oriented processes driven by informational flows instead of predefined control flows.

1.5.6 A Research Road Map for IT-Enabled Services

Research problems related to designing IT-enabled services articulated with the service concept, service systems, and service processes could be summarized as follows:

- How can a systemic thinking be applied to codesign and implement IT-enabled service systems at a large scale through incremental and progressive collaborations among service stakeholders?
- How can this be done by means of adaptive combinations of resources, information technologies, and skills, driven by customers' involvement and requirements and supported by a service-oriented aided infrastructure in dynamic, ad hoc, social-based, and distributed environments?

The foundational logic by which research strategy can be formulated to tackle the main research question of "designing IT-enabled services" should be on the systems thinking theory [41] and the systemic thinking approach [43]. The systems thinking theory guides us to define principles, models, and methods that can be generalized across various service systems, their elements, and the relationship among them [42]. The systemic thinking approach assist us to develop a problem-solving mindset by defining general "patterns," including organizational, business, and technical issues that improve the whole service design process. The challenges related to IT-enabled service systems lead us to organize a research strategy as follows:

1. Designing IT-enabled service systems, focusing on research topics such as service architectural framework, collaborative processes, requirement engineering, service design methods, bundling and ad hoc composition mechanisms, and digital service ecosystem
2. Implementing a service-oriented infrastructure for IT-enabled services, focusing on research topics such as ad hoc Web service composition mechanism with Web service discovery, selection, and information security in social-like, distributed, dynamic, and opened environments

Research contributions and solutions presented in this book focus on IT-enabled services from a system perspective to develop a "service architectural framework" in which IT-enabled systems can be studied, modeled, and designed as combination of resources with respect to "multiple requirements" that are shared by stakeholders from different disciplines and through a "collaborative design method" that gradually builds service processes.

To build large-scale IT-enabled service systems from a business perspective, Chaps. 2 and 3 presents a "design by bundling approach," which simultaneously integrates IT-enabled services by means of "data-driven service processes" to create service bundles. From a technical perspective, Chap. 4 presents a "service-oriented aided infrastructure" to support IT-enabled service systems in ICT-enabled environments with "ad hoc Web service composition mechanism," social-based Web service discovery and selection, and a secured SOA design method.

Inspired by the systemic model that is proposed by Le Moigne [106] to organize information systems into decision, informational and operand subsystems, and the three prerequisites of the service design model that is proposed by Edvardsson and Olsson [47], the design and implementation of IT-enabled service system can be broken down into four subsystems that summarize major contributions discussed in the next chapters:

- **Decision system**: focuses on the service concept to progressively and incrementally capture all service stakeholders' needs and decide on best decisions to cocreate and consume services. At this level, **Chap. 2** includes a service system reference model, service, and customer requirement propagation model for IT-enabled services. A graphical goal-driven model and business rules are also discussed to specify resources in service encounter and collaboration among service stakeholders.
- **Bundling system**: focuses on the service system and relies on service characteristics to assemble existing IT-enabled services and, consequently, build a digital service ecosystem at a large scale. Within IT-enabled service systems, **Chap. 3** presents a collaborative design method and focuses on service processes driven by business artifacts. A business artifact is a business objects, described by states, life cycles, and associated tasks. By using business artifacts to build data-driven processes in services, **Chap. 3** introduces a method to discover business artifacts, identify business artifact interaction patterns, and develop a collaborative design method that handles changes that may occur during service design and provisioning.
- **Operand system**: focuses on a service-oriented aided infrastructure to support IT-enabled service systems. At this level, **Chap. 4** presents a Software-as-a-Service (SaaS) architecture to interact with IT-enabled services (front-offices), an ad hoc Web service composition algorithm based on social-based Web service discovery to build ad hoc composite Web services.
- **AI and IoT enabled service systems**: Artificial Intelligence and data analytics enable services with decision-making supports and insights. In addition, the Internet of Things has created tremendous opportunities for more direct integration of the physical world into the Internet and results in efficiency and economic benefits in various services. **Chapter 5** identifies main research challenges to embed artificial intelligence, predictive analytics, connected devices capabilities, and cybersecurity at different levels of IT-enabled services systems. More precisely, **Chap. 5** presents a multidisciplinary, integrative, and holistic research framework, combining perspectives from service engineering (software engineering), cyber-physical system engineering, cybersecurity, machine learning, and artificial intelligence. The chapter generalizes the concept of Web service model and Web service composition to build smart and secure IT-enabled services with composable AI, privacy-preserving-by-design, and resilient federated learning.

Chapter 2
Service Reference Model and Requirements

Keywords Service modeling frameworks · Service architectural framework · Service design · Service reference model · Service requirements · Service systems · Collaborative service design · Service delivery · Service concept · Modeling services · IT-enabled services · Service system reference model · Ontologies · Business artifacts

2.1 Introduction

As stressed throughout the Introduction, IT-enabled services lack theoretical background and design methods due to a marinade of several factors, including the absence of comprehensive and consistent service definitions [7], the proliferation of services in almost all businesses [107], unconventional characteristics of services [15], and their socio-technical features [24], which appeal insights from multiple disciplines [14]. Moreover, the problem of not having a clear and shared understanding of the service concept has led to service design methods and design processes that are limited to only operational and tactical issues. Current design methods also fail to shift from only improving specific parts in services to focus on the entire service life cycle and optimal strategy to incorporate technologies, processes, people, and physical facilities to create added-value services [108].

Since ICT-enabled environments, which we refer to as a social-based, collaborative, ad hoc, dynamic, opened, and distributed environment, have significant impacts on our perception of the design and development of IT-enabled services, we recognize the relevance of the systems thinking theory and the systemic thinking approach to understand, design, and improve IT-enabled service systems [24, 40, 54, 109].

As a result, ICT-enabled environments and the service concept clearly have key roles to play in IT-enabled service design and implementations, not only as a core construct of the design process but also as a mean of "modeling" services [45]. By service "modeling," we refer to the need for meaningful descriptions (i.e., models) that can be processed by programs and understood by stakeholders from different background to facilitate collaboration in designing and delivering services.

© The Author(s), under exclusive license to Springer Nature Switzerland AG 2023   21
Y. Badr, *Smart Digital Service Ecosystems*, SpringerBriefs in Service Science,
https://doi.org/10.1007/978-3-031-27926-3_2

In the following sections, we elaborate on the service concept and the systemic approach as an important driver for our perception to design IT-enabled services. This concept leads us to identify what are the basic elements (or resources) in IT-enabled service systems and how they can be modeled and used to build service systems and apply a systemic thinking. We particularly focus on the challenges of establishing foundations for service modeling. Moreover, we focus on the service requirements model to "concretize" the service concept by identifying service elements and their representational models. The reference model is a foundational building block for a novel collaborative service design method and a support to develop and bundles IT-enabled services. This chapter also presents an end-to-end requirement engineering approach to capture customer needs at every step in the service design and delivery.

2.2 Services and Challenges

IT-enabled services involve stakeholders from different disciplines, such as designers, IT professionals, engineers, businessmen, developers, regulators, employees, and customers. They collaborate to ensure that service design and provision are focused on satisfying customer needs and aligned with the firm's service strategy. Customer involvement and collaboration are crucial during service design and consumption. Due to service socio-technical characteristics, there are various key barriers to support collaboration among service stakeholders and, consequently, establish a systemic design method for building IT-enabled services as arrangements of technologies, business processes, and people. An important barrier is the lack of a common service concept that is understood by stakeholders. Without a clear and shared understanding of the IT-enabled service to cocreate and deliver (i.e., the service concept), how do managers expect to design a successful service? For even a relatively simple service, a poorly specified or delivered service will easily satisfy customer expectations and fulfill service provider business objectives. Moreover, modeling services still urge a service modeling framework, covering common vocabularies, models, and modeling languages [110] to present and manipulate IT-enabled service systems from different point of views, different perspectives, and different levels [27, 95–97].

2.2.1 The Service Concept

The service concept is frequently used, and most work to date has been concerned with its definition without significant contributions to service design and developments [6]. Edvardsson and Olsson [47] refer to the service concept as the prototype for service and define it as the "*[...] detailed description of what is to be done for the customer (what needs and wishes are to be satisfied) and how this is to be achieved."* The *what* and *how* approach was extended by Goldstein et al. [45], who defined in terms of the customer needs to be satisfied, how needs should be satisfied, what is to be done for the customer, and how this is to be achieved. However, Edvardsson and Olsson employ the service concept as a main driver for service design decisions at strategic, operational, and recovery levels; at the strategic level, the service concept links customer needs and design decisions with the service provider strategic intentions. At the operational level, the service concept becomes particularly useful, since it incorporates service strategy into the design of service delivery systems. At the recovery level, Goldstein et al. argue that embedding the service concept (the *what* and the *how*) into the service recovery design helps to define indicators for evaluating service performances on an ongoing basis and provide feedback loops to recover deteriorated services. At each level, the service concept intervenes in the overall service design and development process.

The absence of a shared service concept that describes how customer needs should be satisfied, and how this is to be achieved leads participants in service design and development to use various vocabularies, models, and methods within the same discipline and across different disciplines [111]. Standard terminologies or service reference models, describing the service concept and system elements, are not yet available to design and develop innovative service solutions [112]. We do not advocate that all stakeholders should expect to use a single body of terminology, but at least they have to agree on general terms (i.e., service system elements or a reference model) shared across all service-related disciplines, unless *de facto* or *de jure* standards are established in the service communities as regards to service models and architectural frameworks. Unfortunately, this is not the case today.

The service concept at the strategic, operational, and recovery levels also overlaps with the general system structure. Le Moigne explains the general system as the conjunction of the structural concepts (e.g., function-structure-evolution) and the cybernetic concepts (e.g., context-structure-goal directness) [106]. In a more mnemonic way, Bertalanffy [42] describes the general system by several actions in an active environment (e.g., spatiotemporal context), characterized by general black boxes (i.e., organized/organizing components) and feedback loops to possibly ensure self-organizing and adaptation. Actions include functions (what is done) and evolution (how to do it) to transform the system itself as well as its environment, with respect to system goals or strategies. The service concept thus matches the "structure" concept in the general system at the intersection of cybernetic and structural concepts. The structure concept mainly includes system elements and their relationships. Identifying these elements becomes crucial to support service strategies (the *what*) and service delivery systems (the *how*). The problem can be summarized as how to identify the service system elements and represent the service concept from different levels (e.g., strategic, operational, recovery), different point of views (e.g., information, collaboration, quality, and infrastructure), and perspectives (e.g., business, computing, social, ...) to support the design of IT-enabled services.

2.2.2 Modeling Services

Before proposing service models for the service concept and agreeing on the structural elements in service systems and set of shared terms to include in the service model, we address the problem of identifying *criteria* upon which we decide whether a term or an element should be included in the service model. According to Weinberg [113], *"A system is a way of looking at the world, it is a point-of-view based on the perception or understanding of a phenomena by one or several observers."* This means that the system perception is "subjective" with respect to an observer point of view and a domain of interest. A view allows an observer to examine a particular set of aspects to find general laws to be used in building the system. Each view provides a representation of the system from a perspective, focusing on specific concerns or criteria. In the system engineering, the viewpoint is a way of looking at a system. Finkelstein et al. [110] argue that viewpoints provide notations, rules, and types of models for constructing a certain kind of view [114]. Thus, a system view is a representation of the system from the perspective of a viewpoint. The term view model (or model for short) is thus related to the view description. As a result, the criteria upon which we have to decide whether a service system element should be modeled and included in the view model are subjective concerns and depend on our perception of 1) how we model the IT-enabled service system and 2) how we define our design method to build IT-enabled service systems. These two points are somehow correlated, and we have to clarify their influence on each other before we continue our discussion on the problem of modeling the service concept and IT-enabled services systems, from different viewpoints and models. Marvin Minsky [115] suggests an interesting observation that relates the designer to the design method and the perceived model: *"To an observer B, an object A* is a model of an object A to the extent that B can use A* to answer questions that interest him about A."* Obviously, this definition asserts that the model is part of the method and has an important role to produce knowledge for the designer by focusing on specific concerns. This implies that the quality of the model depends on its capability to produce useful knowledge. Since the designer (i.e., manager, developer, or casual user) belongs to a specific discipline (i.e., service management), the quality of the model is subject to this discipline and does not depend on other disciplines (i.e., computer science) that help to produce the model (i.e., UML diagrams, service blueprints). In other words, if we adopt methods and models in the computer science as the discipline to design IT-enabled services through collaboration among designers from different disciplines, the quality of the service model should only be measured with respect to all designers and not only to the computer science discipline. This is an important issue in IT-enabled services, which consists of finding a consensus on models accepted by all service stakeholders in a multidisciplinary context, and efficiently used in a "collaborate-to-innovate" design method. Instead of a "one-size-fits-all" view model, different view models should coexist in practice within the same discipline and among different disciplines. The main problem is to ensure the semantic interoperability and establish links and/or transformations between different service view models to facilitate collaboration within the same discipline and among different disciplines and how to correlate these view models and ensure that any changes in one view will be propagated through sibling views (if necessary).

2.2.3 Service Frameworks

Modeling systems from different viewpoints and with multiple models and applying transformation strategies between models have been addressed by several enterprise architectural frameworks to manage information systems [116]. Gartner defines the architectural framework (AF) (a.k.a enterprise architecture framework) as " ... *the process of translating business vision and strategy into effective enterprise change by creating, communicating and improving the key requirements, principles and models that describe the enterprise's future state and enable its evolution [117].*" In practice, the AF helps to discover and execute the best transformation strategies that drive the "enterprise" progress toward desired future states by means of business process integration and business process standards. An AF employs various methods and models to describe the structure and dynamic of an organization from different viewpoints and establish transformation strategies for changing the enterprise. The practice of AF is not new and can be traced to the Zachman enterprise framework in 1987, which is followed by a series of frameworks, such as NIST, POSIX, TAFIM, JTA, JTAA, TOGAF, EAP, C4ISR, TISAF, FEAF, PERA, GERAM, and CIMOSA, just to name a few [118]. In-depth comparative details of these frameworks can be found in the Schekkerman's book [119]. These frameworks aim primarily at:

1. Capturing a strategic vision of the entire organization in all its dimensions and complexity from a holistic perspective (as-is enterprise architecture)
2. Improving business effectiveness and efficiency (to-be enterprise architecture) through transformation strategies (e.g., sequencing plans)

The transformation plan defines the strategy for changing the organization from the current architecture (as-is) to the target (to-be) architecture through a series of actions (i.e., centralization and standardization of business processes). Most of enterprise architecture frameworks produce taxonomies, diagrams, documents, and models to describe the logical organization of business functions, business capabilities, business processes, information resources, software applications, and communications infrastructure. Regardless their maturity, complexity, and transformation strategies, most enterprise architecture frameworks focus on enterprise improvement and evolution by redefining the organization scope and missions, restructuring its business functions, adopting standards, federating business processes, or reengineering business processes with technology alignment to reduce costs. By doing so, they emphasize on the business system itself rather than focusing on customer involvements as a main driver for improving businesses. In his words, Schekkerman states that " ... *most enterprise architecture efforts are too inwardly*

focused, and do not include the customers and key business partners. This results
in miss-aligned enterprise architectures, and lost opportunities to gain competitive
advantage." While enterprise architecture frameworks are significant, in the sense
that they describe architectural artefacts, they do not primarily address service and
collaboration in general neither customer involvements in cocreating business added
value.

Recently, some works tend to enhance architectural frameworks for service
systems [27, 80, 109, 120, 121]; Pineda et al. [109] have studied several architectural
frameworks (e.g., ITIL, e-TOM, TOGAF, NIST) developed for IT services (e.g.,
telecommunications, IT, business reengineering, Web services, etc.). They assert
the absence of unifying frameworks that capture all aspects required to model
service systems. Despite their effort to extend the National Institute of Standards
and Technology (NIST) framework for an end-to-end view of the service system,
they realize the need for integrating tools for real-time dynamic analyses of changes
and variations in service systems. Bicer et al. [121] propose the integrated service
engineering framework (ISE) based on MDD (model-driven development) and
Zachman framework to address requirements of Internet of services and apply the
separation of concerns to model different service dimensions (i.e., processes, actors,
data, and rules) from strategic, conceptual, logical, and technical perspectives.

Architectural frameworks still need in-depth and rigorous research to handle
services and develop different viewpoints and multiple models for designing IT-
enabled services. A strong effort is evident to develop specific frameworks able
to gain understanding on how to model IT-enabled services. However, most existing
architectural frameworks focus on information systems and their business processes.
Recent frameworks emphasize on the service-oriented design from different perspectives and levels. Some of them are too conceptual or technical, focusing on
implementation details (e.g., SOMA [78], CBM [79], SEASIDE [80], ISE [80], and
MRM [81]). They somehow neglect customer involvement in codesigning services
and their active roles in collaboration contexts.

In conclusion, a comprehensive service modeling framework focuses on the
service concept as a driver of designing IT-enabled services, encompasses all
different system elements, and integrates collaboration among service actors to
codesign, codevelop services, and remain an open research challenge, despite
several efforts that have been spent in this direction.

2.3 Design Requirements for Services

As mentioned early, Goldstein et al. [45] employ the service concept as a driver
for service design decisions at *strategic*, *operational*, and *recovery* levels. Similarly,
we employ the service concept at these levels as a driver for our research strategy to
establish foundations for service modeling frameworks for designing and implement
IT-enabled services. At each level, we illustrate hereafter how the service concept

provide guidelines to shape up a potential framework for modeling IT-enabled services.

At the **strategic level**, Goldstein et al. [45] argue that service actors should share a clear understanding of the service concept. In their work, the service concept includes the service strategy of what to deliver and how that strategy should be implemented. By defining the service concept, service providers create an alignment between how they intend to provide and what customers may require or expect. By such, we summarize the following implications/recommendations to establish a service modeling framework:

- A *service system model* should capture what the service provider's design intention and how to share the concept behind the prospective service with clients and participants who are involved in the service design process.
- A *negotiation and trade-off phase* should allow service customers to specify their needs and service providers to specify requirements and propagate these requirements through the service life cycle. Service providers and consumers should be able to incrementally define a collaboration strategy through service design and delivery.
- A *goal-driven model* should support collaboration strategies and adapt to changes through adaptable composition mechanisms to bundle services at large scales (i.e., digital ecosystem) and reorganize resources in response to changes.

These recommendations lead to investigate new research directions for *service system reference models*, *service requirement models,* service bundling, and service composition mechanisms for digital service ecosystems.

At the ***operational level***, Goldstein et al. contend that service providers should organize themselves during design phase to build service delivery systems (*how*) that satisfy customer needs (*what*). Chase and Bowen [92] suggest that the design of a service delivery system includes service participant roles, technology, businesses, and processes by which services are created and delivered. The service concept, thus, incorporates service strategy into the design of service delivery systems. From a research perspective, we settle the following implications/recommendations to establish service modeling frameworks for digital service ecosystems:

- IT-enabled services should be designed and delivered in a problem-solving mindset that foster collaborations to find *service design patterns*, integrating people, technology, and businesses, satisfying customer needs and globally improving IT-enabled services. These patterns postulate reusable solutions to commonly occurring problems.
- *Service processe*s, as supports for service design and delivery, should be driven by business goals and customers satisfaction and enabled with predefined or ad hoc business activities.
- A *service design method* should progressively and incrementally capture service stakeholders' needs and feedback and decide on best decisions to deliver service processes and globally improve their performance.

These recommendations lead to explore new research directions for *service collaboration patterns*, *collaborative design methods*, and *information-driven service processes*, and ad hoc Web service composition.

At the **recovery level**, the gap between provider perceptions of customer needs and the customer expectations leads to disconfirmation and low service quality. A service recovery measures factors to determine low service quality causes (i.e., KPI, data analysis, correlational studies) and establish scenario to improve service quality. Goldstein et al. argue that embedding the service concept (the *what* and the *how*) within service recovery design provides a framework for continuously evaluating service qualities. From a research perspective, we settle the following aspects to integrate service recovery in designing services:

- ICT-enabled environments should implement qualitative and quantitative performance indicators to measure service quality during design and consumption.
- ICT infrastructures should support IT-enabled services and provide feedback loops to measure gaps between what customers expect and providers perceive.

These recommendations lead to investigate mechanisms to capture *service quality* in ICT infrastructures such Software-as-a-Service (SaaS) and service-oriented architecture.

2.4 An Architectural Framework for IT-Enabled Services

Since the service concept supports the service design decisions at *strategic*, *operational*, and *recovery* levels, we developed several contributions for a IT-enabled service architecture framework for IT-enabled services, including two main models: the *service system reference model* [122], which supports a generic service system for knowledge-intensive services [123], and the *shared requirement model* [124, 125], which specify requirements from service consumers and service providers in knowledge-intensive firms [126]. These models provide general guidelines for establishing service architectural framework for extending Web service composition [124] and multidimensional service ecosystems [127]. The *service system reference model* comprehends all aspects of service system elements from different viewpoints and aims to concretize or represent the service concept from several view models: systematic view, characteristic view, business view, and interaction's view, each of which is support by an ontology-based model. All views, describing service system elements, are interrelated by means of ontologies, providing thus a backbone for elaborating complex models and methods such as the requirement model and the collaboration model, which we use in our proposed collaborative design method. The *shared requirement model* refers to views in the service system reference model to specify unclear and imprecise customer needs and specify them with requirements to propagate through the service life cycle. It includes the goal-oriented graphical model, the service requirement modeling language (SRML), and the mapping algorithm to generate from requirements business objects and rules.

The originality of these contributions relies on ontologies to "concretize" the nature of the service concept and the service system structure by creating multiple view models, describing different viewpoints. These views are not isolated from one another, but they are interrelated by means of ontologies. Ontologies are organized into three levels: the upper ontologies, which are refined by the generalized ontologies, which are also refined by the specialized ontologies. Our set of view models and their graphical representations can be easily shared by stakeholders from different background during collaboration and can be easily processed by computers.

2.4.1   Service System Reference Model

The *service system reference model* provides different views of the service system, namely, the *system view*, the *characteristics view*, the *business view*, and the *interaction view*. The reference model ultimately aims at showing how to gain a common understanding on fundamental service system elements and their relationships. As a reference model, it embodies basic elements that exist in service systems.

As depicted in Fig. 2.1, the service system reference model consists of different views to present the service system from different perspectives by focusing on specific concerns. To model these views, a set of ontologies are used to provide formal descriptions of concepts, relationships, restrictions, and axioms related to service system elements. Ontologies define a common vocabulary that can be shared by various service participants from different disciplines and provide a

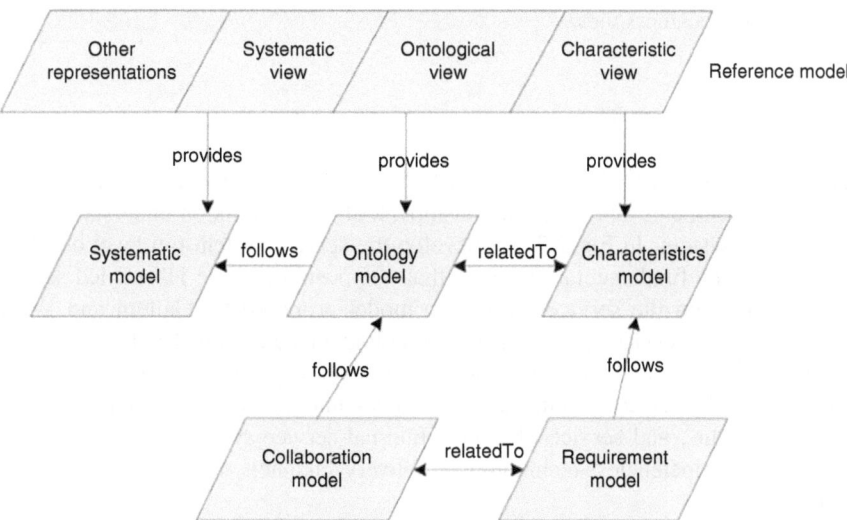

Fig. 2.1 Service system views and models in the service modeling framework

common understanding on how to deal with ambiguities during collaboration. Another important issue is that ontologies, as a modeling language, are the "glue" that connects all service system views and makes possible the mapping between concepts between views and the propagation of changes. In the service system reference model, each view can be defined as a successive refinement of three types of ontologies:

- **Top-level ontologies**: build one-size-fits-all ontologies to describe generic service system concepts. For example, the top-level ontology of "educational service systems" can be described by generic concepts, such as resources, competencies, technologies, delivery channels and service quality, and customer inputs.
- **Generalized ontologies**: refine top-level ontologies with additional and/or specialized concepts to build domain-focused ontologies. For example, the generalized ontology of the "e-learning services" is a refinement of the top-level ontology of the "educational services."
- **Specialized ontologies**: refine generalized ontologies with additional and/or specialized concepts to build custom-tailored ontologies. For example, the specialized ontology of "ambient assisted learning services" is a refinement if the generalized ontology of "e-learning services" with a particular focus on problem-solving and innovative context related to a particular customer.

As a result, the ontology-based views support the service system reference model with computational and reasoning capabilities. They particularly facilitate collaboration among various service actors and ensure semantic interoperability between view models. Keep in mind that the service system reference model is extensible and new views can be added to describe the service system from new perspectives and viewpoints (e.g., quality view, value view, etc.). In the following sections, we briefly present four views: the system view, the characteristics view, the interaction view, and the business view.

2.4.1.1 The System View

The system view acts like a blueprint and describes the service system in terms of its basic elements and their relationships. Il allows a common understanding on the service system. In Fig. 2.2, we develop the service system top-level ontology to describe the fundamental elements that compose a generic IT-enabled service system based on the service innovation model proposed by Gallouj and Weinstein [128]. The service system top-level ontology includes three levels: component level, system level, and the system-of-systems level in accordance with the systems theory [41]. A service system mainly involves internal service system elements, customer inputs, and service offerings. Internal service system elements include resources, competencies, technologies, delivery channels, and service actors as follows:

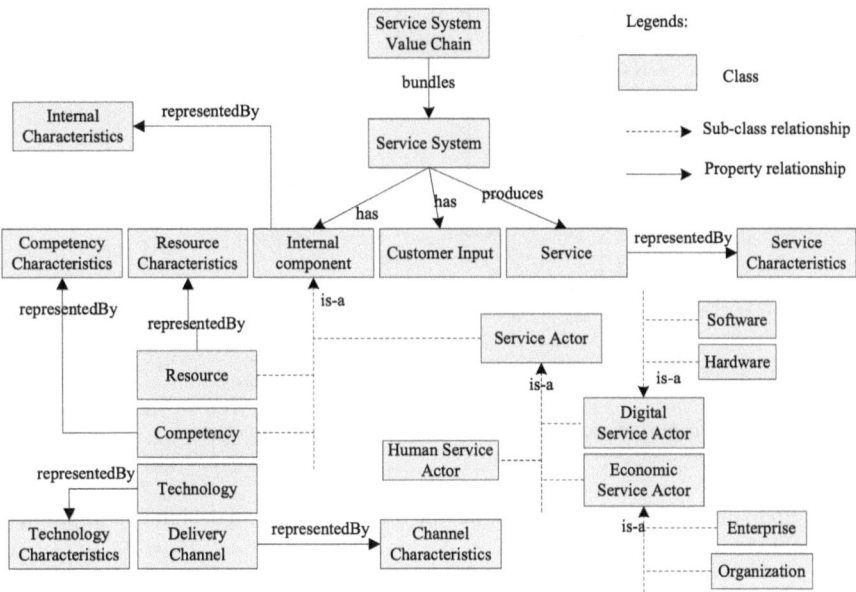

Fig. 2.2 The system view based on the service system top-level ontology

- **Resources** are assets from which benefits are produced, such as computing resources, capital resources, intangible resource, and physical resources.
- **Competency** is the ability to perform a specific role and do a job properly. It refers to creative competency, relational competency, strategic competency, learning competency, and management competency.
- **Technology** refers to a collection of technical methods, skills, processes, techniques, and tools to solve a problem or achieve a goal. It includes scientific technology, ICT, organizational technology, etc.
- **Delivery channels** provide means for communications. They consist of organizational channels, physical and electronic channels, etc.
- **Customer input** includes customers themselves, knowledge, and assets.
- **Service actors** refer to services or service stakeholders participating in the service design and delivery.

In this view, the provision of any service can be described in terms of a set of its final characteristics that reflect its external properties seen from customer's point of view. Based on the top-level ontology, we introduce a simple presentation of the service system as a combination of internal components grouped into vectors, including competences [C], technology [T], resources [R], and delivery channels [D] (Fig. 2.3). The provision of final characteristics mobilizes customer and provider competences to exploit resources and combine different technologies and collaborate through delivery channels to exchange resources. The main value of the systematic view is its ability to provide a static view of service system elements

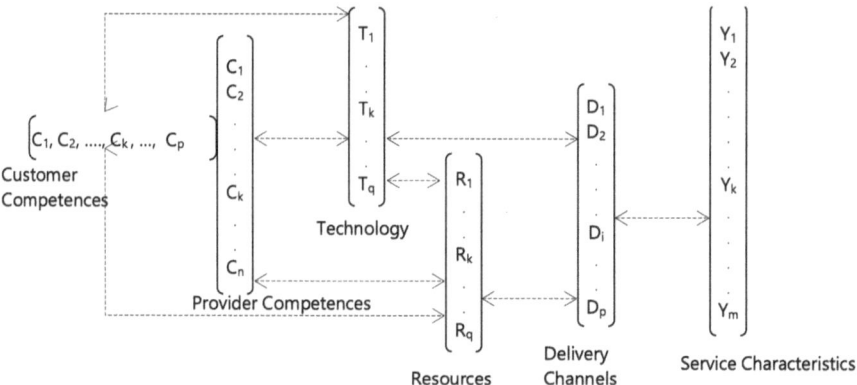

Fig. 2.3 The IT-enabled service as a combination of internal components

and help to define design patterns to decide on best actions to design and implement IT-enabled services.

2.4.1.2 The Characteristics View

The characteristic view to describe qualitatively service system components as a set of characteristics. A characteristic corresponds to a service quality or attributes such as benefits, capabilities, abilities, and behavioral or mental states. Service characteristics are exposed to customers to perceive the benefit from the service, whereas internal characteristics refer to service system components and mainly intend to hide their complexities to facilitate collaboration among service stakeholders from different disciplines. Each characteristic can further be measured and annotated by a set of features exploitable. This view can be extended with functional utilities to describe a set of actions a customer can perform to receive benefits from the service system. Based on characteristics, features and functional utilities IT-enabled service can be identified to create a service bundle or customized to satisfy customer needs. The service system thus mobilizes its internal elements to support its exposed characteristics (Fig. 2.4).

Service characteristics were coined to Lancaster [129], who argues that both tangible goods and intangible services can be described by a set of characteristics that a good/service embodies. Gallouj and Weinstein [128] consider that services can be represented by sets of final characteristics, which can be achieved by service actors using their competencies and technologies.

The example of the online banking service in Table 2.1 is presented to clients as a vector of characteristics. Characteristics such as price, speed, and dependability can be observed and measured, whereas characteristics such as reputation, quality, and safety are qualification criteria and cannot be directly measured.

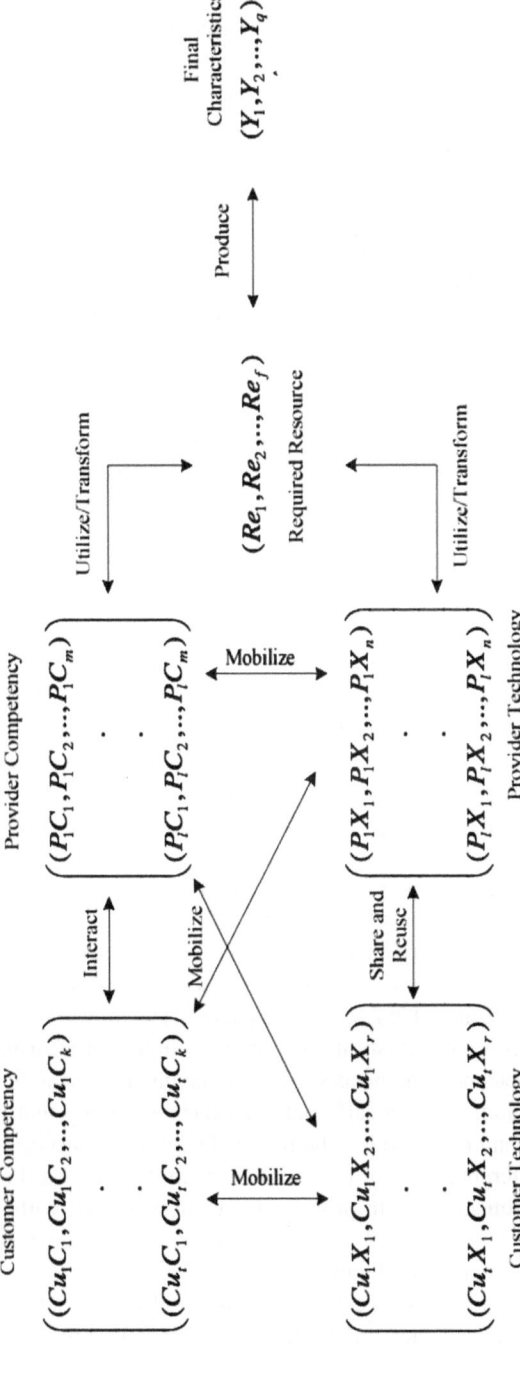

Fig. 2.4 A service system characteristics view

Table 2.1 Examples of final characteristics

Characteristics	Characteristics
Convenience, e.g., site location	Quality, e.g., perceptions important
Dependability, e.g., on-time performance	Reputation, e.g., word-of-mouth
Personalization, e.g., know customer's name	Safety, e.g., customer well-being
Price, e.g., quality surrogate	Speed, e.g., avoid excessive waiting

The main value of this view is that "thinking" about service system elements in terms of their characteristics makes service actors focus on their core businesses and their interactions rather than attempting to understand specific domain or technical terms related to different disciplines. Describing IT-enabled services as sets of qualitative and quantitative characteristics and constraints helps to improve the service quality at design time and define indicators to service performance at run time.

The characteristics view is mainly used in the service design collaborative (Chap. 3) and service bundling and composition (Chap. 4) and makes possible to separate internal system elements from external service utilities [77].

2.4.1.3 The Business View

The business view handles explicit business knowledge in IT-enabled services. In this context, a business knowledge is a representation of a business concept identified by its name, purpose, attributes, behavior, relationships, and constraints. Business knowledge represents, for example, a transcript, invoice, or customer along their life cycles. Business knowledge embodies business logic, facts, and constraints that show how a business concept can achieve a business goal along its life cycle (e.g., creation, manipulation, archiving...). Sharing business knowledge facilitates interactions with customers, keeps them informed, and helps make choices during service design and provision [130].

Since the design and implementation of IT-enabled services are driven by business knowledge instead of business processes (workflow of activities), we strategically decide to model business knowledge with *business artifacts* [131]. A business artifact is a mechanism used to record chunks of information that can be easily used by business managers without having technical backgrounds in information management. It is a self-contained business entity that includes a set of attributes, states, and a life cycle. The life cycle describes successive changes in states, reflecting different stages of the artifact manipulation toward its final states. The life cycle also reflects how to process its business artifact without explaining how to do it. It is a "road map" with memory capturing the business artifact evolution, from creation to completion.

Due to the importance of business artifacts in building IT-enabled services and digital service ecosystems, the order business artifact in Fig. 2.5 is a concrete

Informational Model	Transition rules	
Attributes	**R1: If** DEF(prodName) ∧ DEF(prodType) ∧ DEF(bid) **Invoke** Estimate	
ProdName: string	**R2: If** approved = **true** ∨ execApproved = **true Move to** Fulfilled	
prodType: string	Tasks	Lifecycle
bid: interger	**Estimate**: profitMargin	Created
profitMarging: [0..10]	PRE: DEF(prodName) ∧ DEF(prodType) ∧ DEF(bid)	
scheduleDate: date	EFF: Bid ≤ 400 → rofitMargin ≤ 25%	Updated
States	**RoutineApproval**: approved	
Created: boolean	PRE: DEF(bid) ∧ DEF(profitMargin)	Approved
Updated: boolean	EFF: Bid ≤ 100 → approved = *true*	
Approved: boolean		Fulfilled
Fulfilled: boolean	**ExecApproval**: Approved	
Archived: boolean	PRE: approved = *false*	Archived
	EFF: *t*rue → DEF(Approved)	

Fig. 2.5 The order's business artifact (source [132])

example of the business artifact, which is defined by a set of attributes, states, life cycle, two transition rules, and three tasks. Each rule is specified by a formal logical expression with two parts: condition and action (C/A). If the condition holds, the action is applied. Conditions are constraints on states, attribute definitions (DEF), and their values. Possible actions include transition rules that execute tasks, which may change attribute values (e.g., R1) or states (e.g., R2) in accordance with the life cycle, which is represented as a directed graph in Fig. 2.5. An artifact task is a declarative and intentional specification of the business artifact's behavior, describing*preconditions* to be satisfied in order to produce effects on business artifact attributes and states after the execution of a task. A typical rule to describe the intentional statement has the following pseudo-code syntax: If the *precondition* holds, then the effect is evaluated. The execution of each task requires that preconditions hold by checking current states and whether attributes are defined (DEF) or having specific values. The rule updates states and attributes as specified in its post-condition.

In the business view, the top-level business artifact ontology, depicted in Fig. 2.6, shows the artifact representation in terms of attributes, states, and life cycle concepts and describes exchanged knowledge with structured, unstructured, and semi-structured informational representations. It is worth noting that business artifacts can be processed manually or automatically by software applications or Web services. The main value of the business artifact modeling approach in IT-enabled services can be summarized from different perspectives as follows:

• Business artifacts are the basis for the factorization of business knowledge, because they are self-contained business records, embodied with states (facts) and life cycles (skills), capturing their manipulation from creation to completion.

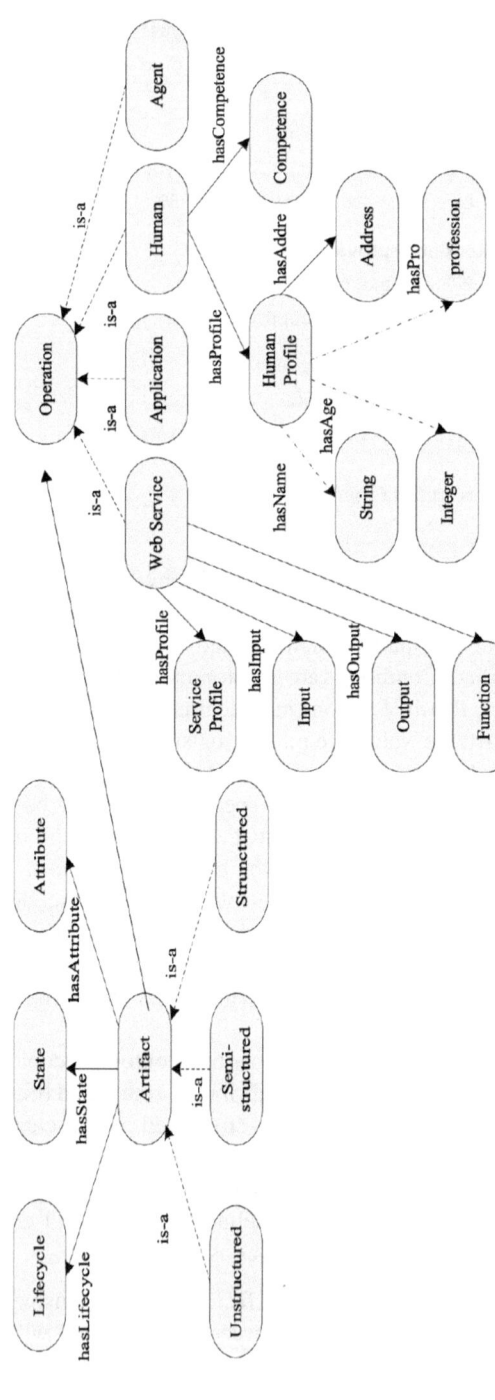

Fig. 2.6 Business artifact top-level ontology

- Business artifacts are key information pertinent to a business context. They are familiar construct to the business community and manageable, analyzable, and flexible from the perspective of business specialists.
- Business artifacts make a clear separation between business data and software applications that manipulate them. The logic of a business artifact processing and interactions among business artifacts are only specified with transition rules and thus are not embedded in software.

Modeling business knowledge with business artifacts has thoughtful implications on the overall approach to design and implement IT-enabled services. As described through our chapters, business artifacts are fundamental constructs in many of our contributions, namely, the requirement model, the collaboration model, the collaborative design method, and the service bundling. As opposed to the activity-driven modeling approach (i.e., business processes), business artifacts support data-driven modeling approach and techniques to build service processes.

2.4.1.4 The Interaction View

The interaction view focuses on interactions between existing IT-enabled services, customers and IT-enabled services, or collaboration between stakeholders. The interaction view adopts the "emergent architecture" style to deal with the growing variety and complexity of IT-enabled service interactions. The "emergent architecture" is an enterprise architecture recommended by Gartner, which can be summarized as " [...] 'architect the lines, not the boxes', which means managing the connections between different parts of the business rather than the actual parts of the business themselves [117]." This architectural trend incites to model relationships between systems as interactions via some set of formal, informal, or manual interfaces. This view supports the bundling process and establishes potential interactions between existing service systems in order to bundle them into a new service system, satisfying customer needs [133]. In this context, the "interfaces" refer to the final (exposed) characteristics (e.g., functional utilities), "lines" denote the delivery channels, and the "boxes" refer to the internal service system elements or the entire service systems. The interaction view thus provides a general strategy to integrate two (or more) service systems or building digital ecosystems of IT-enabled services (Fig. 2.7).

The main value of the interaction view is to provide bundling, as an alternative approach, to create new service systems from existing service systems at a large scale based on service final characteristics instead of the collaboration approach, which seeks to cocreate new services from scratch. As illustrated in Chap. 4, the service collaboration approach describes how to specify and match customer needs (expected service characteristics) with respect to the exposed service characteristics to assist service providers and customers to cocreate new services. The Interaction view describes how to bundle services with data-driven processes and provides a Software-as-a-Service – model for customers to interact with IT-enabled services.

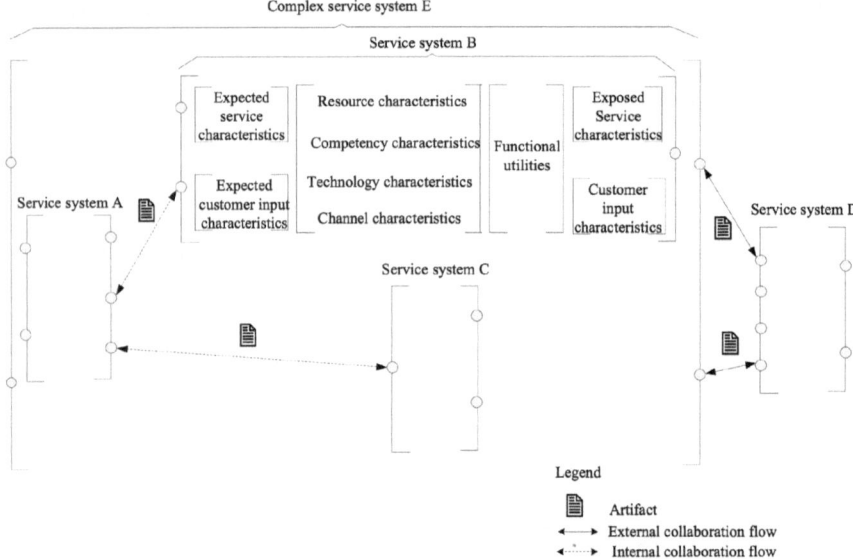

Fig. 2.7 Bundling service system by comparing service final characteristics

2.4.2   The Service Shared Requirement Model

Requirements engineering (RE) is the process of discovering the purpose for which
a system is intended by identifying and documenting customers and their needs
in a form that is amenable to analysis, communication, and implementation [134].
Requirement engineering receives a lot of attentions in manufacturing, software
development, and construction projects. Most of service requirement approaches
involve customers only in the preliminary phases and produce technical specifications related to a specific discipline (e.g., software development) [135, 136].

Building IT-enabled service systems requires a novel approach to identify and
specify requirements when stakeholders from different disciplines collaborate at
each phase of the service life cycle. Few works attempt to extend the requirement
engineering to handle different situations [99, 137, 138]. Nevertheless, they are not
relevant for IT-enabled services, where multi-stakeholders collaborate to specify IT,
businesses, technical, and organizational requirements with different models and
languages to progressively design and develop services [139].

The service shared requirement model depicted in Fig. 2.8 aims to capture
customer needs with a well-defined service requirements expressed as an arrangement of *competencies*, *resources*, and *technologies*. To this end, the requirement
acquisitions and specifications rely on negotiations and trade-offs between customers' needs and service provider capabilities. Requirements thus have to undergo
several phases, including initial contact, negotiation, agreement, and supervision of
expected services and involving informal and formal languages as well as natural

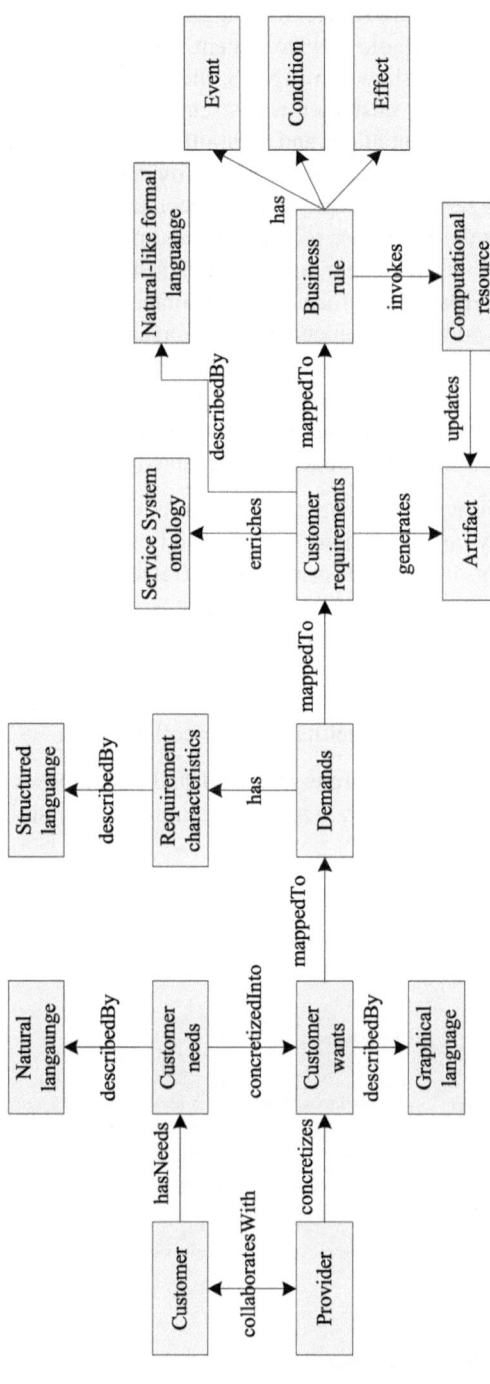

Fig. 2.8 The requirement model top-level ontology

and technical languages. Another important aspect of requirements in IT-enabled services is that requirements are subject to changes and their change effects should be propagated through the whole service system, since the service is design and delivery simultaneously. As shown in Fig. 2.8, the requirement model consists of four levels: customer needs, customer wants, customer demands, and customer requirements, illustrating negotiations and trade-offs between customers and service providers to address drawbacks due to service providers and consumers' misunderstanding regarding service perception and expectations and linguistic barriers. Figure 2.8 also shows that customer requirement and provides insights to identify business artifacts and their business rules. In this sense, the requirement model refers to the business view. It also focuses on customer involvement, and handling changes to ensure top-down requirement propagation) and enabling collaboration to facilitate the communication between service actors from the same discipline (i.e., sideway requirements). The requirement model simultaneously involves a top-down and sideway requirement engineering approaches to jointly allow various service actors from different disciplines to codesign IT-enabled services.

- The *top-down requirement propagation* aims at concretizing unclear customer needs expressed by customers into specified and clear customer requirements through negotiation. It is a goal-oriented requirements approach that helps to propagate customer requirements from upstream service actors (i.e., customers, administrative officers, and business analysts) to downstream service actors.
- The *sideway requirements propagation* aims at representing requirements in a common structured language understandable by all service stakeholders belonging to the same discipline and facilitating their collaboration.

Figure 2.9 illustrates the requirement model and decomposes it into various levels ranging from *want*, *need*, *demand* to *requirement* in a similar way to the

Req. model	Tools	Actors	Characteristics
Needs	Natural language	Customers	•Easiness •Powerful expressiveness •Lack of semantics
Wants	i*model	Customers requirement engineers IT specialists, …	•Graphical interface •Easy to understand •Limited expressiveness •Lack of semantics
Demands	Service Requirement Modeling Language (SRML)	Customers, requirement engineers IT specialists, …	•Formal language •Fit to describe service characteristics •Limited expressiveness
Requirements	Semantic Business Vocabulary and Rule Language (SBVR)	Customers requirement engineers business experts IT specialists, …	•Formal core but natural-like interface •Possible to map to other implementation languages •Not directly processed by software
Technical specification	Web Ontology Language (OWL) & Semantic Web Rule Language (SWRL)	Customers, IT specialists, developers, …	•Formal •Easy to write •Perfect integration with OWL •Limited expressiveness

Fig. 2.9 Multi-layer and shared requirement model

requirement hierarchy proposed by Baida in his work on service bundling [77]. Herein, customer needs expressed in the natural language are first concretized to customer wants by defining expected service characteristics as goals to achieve during collaboration. Customer wants are then specified into customer demands, which specify features and constraints on service characteristics and what are the service system elements that should be combined. Finally, customer requirements are obtained from customer demands. Customer requirements are expressed with business rules and vocabulary and defined with structured English language easy to understand by all service stakeholders. Figure 2.9 shows the purpose of each level and list specification tools or languages that we have developed, and we briefly explain in the next sections.

To describe how service requirements are derived from needs, wants, and demands in the requirement model (Fig. 2.9), the following building blocks are used:

1. The *goal-oriented graphical model* based on i* to express customer wants
2. The *service requirement modeling language* to express customer demands
3. The *mapping algorithm* to generate from SBVR rules, describing customer requirements, a set of business artifacts, and their rules in SWRL

2.4.2.1   Specifying Requirements with Service Characteristics

By referring to the characteristics view, customers set requirements on the (structured) service characteristics, thereby obtaining "expected characteristics" (i.e., customer requirements) through negotiation and trade-offs with the service provider. Since the expected characteristics are related to service characteristics, which are related to the internal service system components characteristics, any change in customer requirements will be propagated, and the service system will remobilize its internal components to satisfy the new requirements (see Fig. 2.10).

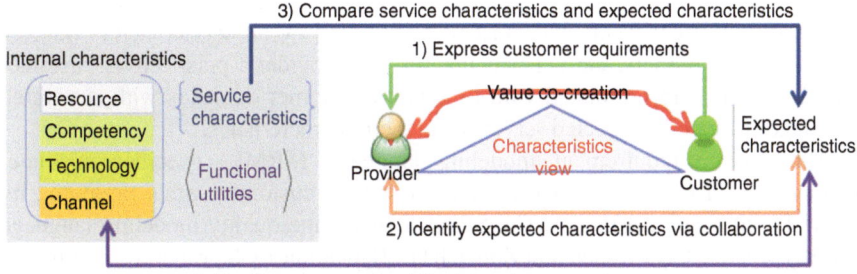

Fig. 2.10 Mapping requirements to service characteristics

2.4.2.2 The Goal-Oriented Graphical Model

The *goal-oriented model* graphically captures the service concept and links it to the requirement model and the system view. It initiates the requirements' assessment through collaboration among customers, requirements engineers, and IT-specialists to express customers "*needs*" in terms of set of goals (soft goals) and decompose them into customers "wants" defined as subgoals (hard goals) to be achieved. The graphical model extends the graphical *i* framework* [140] and is based on the Donzelli [137] software requirements engineering framework, which includes a goal modeling phase and an organization phase. In the goal modeling phase, designers refine soft goals into hard goals, considering quality features, context, and functions. A soft goal can be decomposed into a set of achievable hard goals, tasks, and constraints. During the organization phase, designers use information obtained from the goal modeling phase and elaborate an organization model for each hard goal, including agents (i.e., human), tasks (i.e., well-specified activities), resources, and organizational contexts. Soft goals and hard goals are dependency links, describing the relationships among them. By considering customer "wants" as soft goals representing expected service characteristics, service customers and providers collaborate with each other to transform the soft-goals into a set of hard goals to achieve service characteristics by setting constraints on possible combinations of service system components. The main value of the graphical model is its simplicity. It is quite effective to describe the service concept with customer as a set of goals (what) and select the service system components that should be combined together (how) (Fig. 2.11).

2.4.2.3 The Service Requirement Modeling Language

Based on the goal-oriented graphical model, service providers negotiate and agree with the customers on structured descriptions of their "wants" to avoid ambiguity and to be used as a basis for their service agreements. Pohl [141] argues that customer requirements should be represented in a common agreed formal language to specify customer wants and must be understood by all service actors (i.e., IT specialists, developers, etc.). From the service provider's perspective, customer wants should be reformulated with to express customer demands with structured languages to specify expected services and their characteristics.

The service requirement modeling language (SRML) specifies customer demands in IT-enabled service systems based on customer wants presented by the i*-based graphical model. The SRML extends the quality modeling language (QML) proposed by Frolund and Koistien [142] to define QoS properties in software engineering to rigorously. Conversely, the SRML is used by business experts, IT specialists, and developers to describe customer requirements in terms of service characteristics, service features, and constraints on service system components. The syntactic grammar of the SRML is illustrated in Fig. 2.12. A service characteristic is described with various features, each of which has a type, a scale unit to

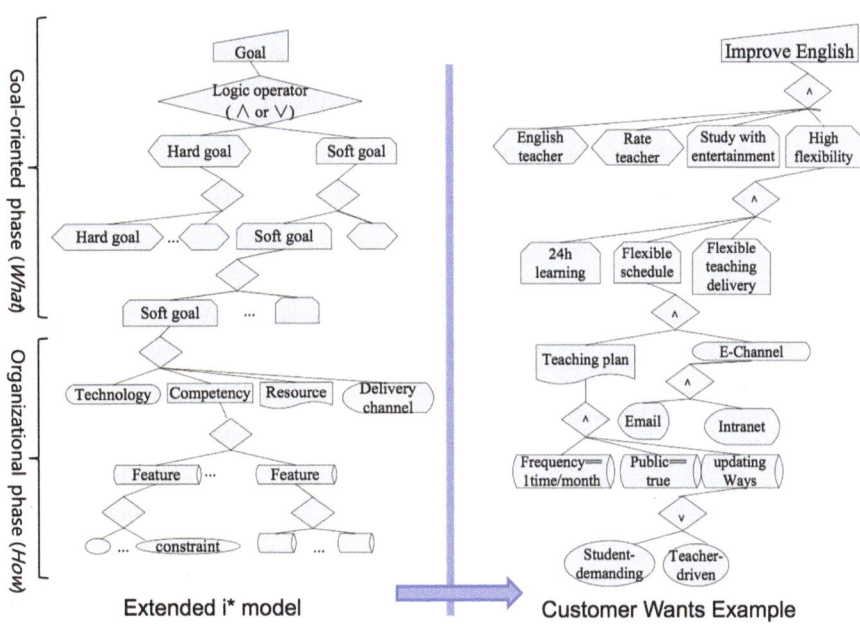

Fig. 2.11 The i*-based model specifying "needs" and "wants"

measure value (i.e., ordinal, interval, nominal, and ratio), value order, and value sequence (decreasing or increasing). Since service characteristics often describe the IT-enabled service qualitatively to help customers to understand the service concept behind the service, we associate a set of features as quantitative attributes to provide measurement metrics. The main value of the SRML is its capability to specify quantitative and qualitative constraints on IT-enabled service system elements (e.g., service characteristics, system components, ...) using a structured language particularly understandable by IT specialists and developers and could be exploitable by software

2.4.2.4 Customer Requirements with SBVR-Based Templates

Customer requirements specify service capabilities, expected characteristics, or qualities of the expected service. Customer requirements should also allow service actors to define business rules to guide the design and implementation of IT-enabled services. The specification of customer requirements is not an easy task, because it should not only be easily understood by customers but also all service stakeholders and translatable into other representations (i.e., technical requirements). The SRML syntax is quite technical and difficult to be understood by customers. For these reasons, we adopt the semantics of business vocabulary and business rules (SBVR) [143]as a basis for formal natural language declarative descriptions

```
expectedCharacteristics ::= ExpectedCharacteristics
{constraint₁;constraintRelationOp constraint₂; constraintRelationOp ...;
constraintRelationOp constraintₖ;}
constrain ::= featureName constraintOp constraintValue
            | featureName{aspect₁; && aspect₂;&&...;&&aspectₖ;}
constraintRelationOp ::= && | ||
constraintOp ::= == | >= | <= | < | > | ! =
constraintValue ::= literal valueUnit | literal
literal ::= string | {n₁,...,nₖ} | number | bool
valueUnit ::= valueUnit/valueUnit | % |valueUnit | Mbits | s | year |hour|
...
bool ::= true | false
aspect ::= percentile percentNum constraintOp constraintValue
        | mean constraintOp constraintValue
        | variance constraintOp constraintValue
        | frequency freqRange constraintOp number%
freqRange ::= constraintValue
            | lRangeLimit constraintValue, constraintValue rRangeLimit
lRangeLimit ::= ( | [
rRangeLimit ::= ) | ]
percentNum ::= 0 | 1 | ... | 99 | 100
characteristicsType ::= InternalCharacteristics
{featureName₁;featureType₁;...; featureNameₖ:featureTypeₖ;}
 InternalCharacteristics::=SupportintInputCharacteristics |
CustomerInputCharacteristics
featureName ::= string featureScale
featureScale ::= ordinal | interval | nominal | ratio
featureType ::= featureSort | featureSort unit
SupportintInputCharacteristics ::= ResourceCharacteristics |
CompetencyCharacteristics | ChannelCharacteristics |
TechnologyCharacteristics
CustomerInputCharacteristics ::= CustomerResourceCharacteristics |
CustomerCompetencyCharacteristics |
CustomerChannelCharacteristics |
CustomerTechnologyCharacteristics
featureSort ::= enum {n₁,...,nₖ}
             | relSem enum {n₁,...,nₖ} with order
             | set { n₁,...,nₖ }
             | relSem set {n₁,...,nₖ}
             | relSem set {n₁,...,nₖ} with order
             | relSem numeric
             | boolean
             | string
order ::= order {nᵢ<nⱼ,...,nₖ<nₘ}
valueUnit ::= valueUnit/valueUnit | % | valueUnit | Mbits| s | year | hour
|...
relSem ::= decreasing | increasing

....
```

Fig. 2.12 The service requirement modeling language

to express customer requirements. The SBVR offers a vocabulary for describing meaning and is an integral part of the OMG's model-driven architecture (MDA). It is intended to formalize complex rules (e.g., operational rules, security policy, regulatory compliance rules) and interpreted and used by computer systems. The following example is a SBVR rule defined with noun concepts (the equivalent of OWL classes), fact types (i.e., the equivalent of OWL relationships), and constraints and cardinalities.

modal operator   noun verb   quantifier   noun qualifier noun

It is obligatory that each order *has* no more than one client *of* Swiss Bank

Based on SVBR provides many advantages. From customer perspective, the SBVR provides structured natural language vocabularies that both users and business analysts can use to express rich business terms and business rules. SBVR distinguish the meaning of business rules and vocabularies from their expressions in textual forms, which indicates that it supports multilingual development. From a technical perspective, the SBVR can be mapped to ontologies, which are the core concept in our service system reference model.

Figure 2.13 illustrates how SBVR is used to facilitate collaboration among customers, business experts, IT specialists, and developers. At the business level, customer requirements are described in SBVR and validated by service customers and providers. At the technical level, the SBVR based-vocabularies in requirements are transformed into new OWL concepts, enriching the service system ontologies (e.g., system view) and business artifact ontologies (e.g., business view). The SBVR-based rules are mapped into SWRL rules to build transition rules triggering tasks to manipulate generated business artifacts.

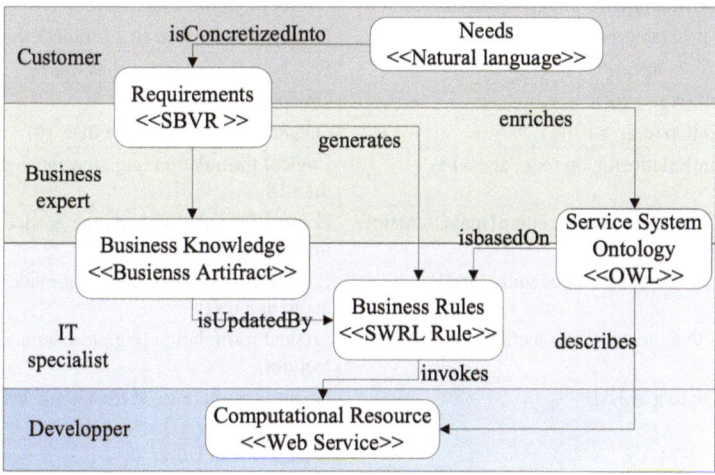

Fig. 2.13 The requirement model viewed by service actors

2.4.2.5   Requirement Mapping Algorithms

The Mapping Algorithm from SRML to SBVR To assist service participants to specify customer requirements, we have developed a rewriting algorithm to generate SBVR vocabularies and rules and facts from customer demands expressed with SRML.

Table 2.2 briefly illustrates a subset of rules that guides our algorithm to map SRML concepts to SBVR concepts with examples from an e-learning case study.

The Mapping Algorithm from SBVR to Business Artifacts It aims at extracting intrinsic knowledge embedded in the SBVR vocabulary and SBVR rules to enrich the service system ontologies and business artifacts ontology with new business artifacts and SWRL rules that manipulate them by invoking software applications. The Semantic Web Rule Language (SWRL) express hornlike rules and logic combining both the OWL DL and OWL Lite with a subset of the Rule Markup Language [144]. Each SWRL rule is of the form of an implication between an antecedent and consequent. The antecedent is a condition over artifact attributes and states, and the consequent is an applied action to manipulate business artifacts. This algorithm translates SBVR concepts such as noun, name, verb, quantification, logical formulation, and modal operation to OWL axioms. For the sake of brevity, we only present the rewritten rules with functional-style syntaxes in Fig. 2.14.

The Mapping Algorithm from SBVR to Web Services The mapping from SBVR to business artifacts is completed with another mapping from SBVR to Web services

Table 2.2 Rewriting rules from SRML to SBVR

SRML concepts	SBVR concepts
CharacteristicsType (e.g., humanResource)	Object (e.g., humanResource)
ExpectedCharacteristics (e.g., student, Jim)	Object or individual (e.g., student, Jim)
Feature (e.g., age, title)	Fact type (e.g., has age, has title)
valueUnit (e.g., year)	Object (e.g., year)
ConstraintOp (e.g., >10)	Quantification (e.g., more than 10)
ConstraintRelatioshipOp (e.g., age >18)	Logical formulation (e.g., the age is greater than 18)
Feature type (e.g., gender: enum{male, female})	Logical formulation (e.g., the gender is male or female)
set{ ,..., } (e.g., set{video, audio, text})	Logical formulation 'and' (e.g., video and audio and text)
charaOp (e.g., teacher && lecturer	Logical formulation (e.g., teacher and lecturer)
freqRange (e.g., [3,5])	Quantification (e.g., at least 4 and less than 5)
enum{ ,...', }	Logical formulation 'or'
Service characteristics	SBVR expressions

SBVR	OWL 2 DL	SBVR	OWL 2 DL
Noun	Class	m is p or q but not both	ObjectIntersectionOf(ObjectUnionOf(:p :q) ObjectComplementOf(ObjectIntersectionOf(:p :q)) Or DisjointUnion(:m :p :q)
Name	individual	If p then q	SubClass(:p:q)
Verb	ObjectSomeValuesFrom(), ObjectAllValuesFrom(), DataSomeValuesFrom() or DataAllValuesFrom()	p if q	SubClassOf(:q :p)
Each	SubClassOf	p if and only if q	EquivalentClass(:p :q)
At least one	objectMinCardinality or dataMinCardinality	Not both p and q	ObjectUnionOf(ObjectIntersectionOf(:p ObjectComplementOf(:q))ObjectIntersectionOf(:q ObjectComplementOf(:p)))
At least n	objectMinCardinality or dataMinCardinality	Neither p nor q	ObjectUnionOf(ObjectComplementOf(:p) ObjectComplementOf(:q))
At most one	ObjectMaxCardinality or dataMaxCardinality	p whether or not q	p
At most n	ObjectMaxCardinality or dataMaxCardinality	At least n and at most m	objectMinCardinality and ObjectMaxCardinality or dataMaxCardinality and dataMinCardinality
Exactly one	FunctionalDataProperty or FunctionalObjectProperty	p and q	ObjectIntersectionOf(:p :q)
Exactly n	ObjectExactCardinality or dataExactCardinality	p or q	ObjectUnionOf(:p :q)
More than one	objectMinCardinality or dataMinCardinality	It is not the case that p	ObjectComplementOf (:p)

Fig. 2.14 Rewriting rules from SBVR to OWL

(i.e., software applications). SBVR proffers service providers and customers a structured natural language to express what-to-do to achieve their business objectives with high-level business requirements. Web services are software applications that specify how-to-do actions to achieve business tasks [145]. Current Web service composition languages that specify business processes are technical or formal languages and fail to be expressive enough to capture business objectives [105]. The mapping algorithm aims to reduce the gap between high-level requirements, describing business objectives and Web service composition requirements.

2.5 Concluding Remarks

Modeling complex systems from different views to focus on specific concerns is not a new approach. It is a common practice in architectural frameworks [119]. Most architectural frameworks dedicated to service systems are mainly available in conference proceedings without a comprehensive list of service elements and without practical design methods to build IT-enabled services [27, 96, 97]. However, the service modeling framework enabled by the reference and requirement models for designing and implementing IT-enabled services differ from traditional architectural frameworks in different ways. For example, the ontology-based specification provides a common formalism to uniformly describe concepts and their relationships in each view and interrelate various views to ensure that changes in one view will be propagated through related views. Ontologies, which formally represent structural containers for organizing information, also support our research strategy to build IT-enabled services, as information-driven processes instead of activity-based processes driven by predefined control flows. The successive refinement of these ontologies allows service stakeholders to collaborate with an appropriate level of details that interest them. Without being an exhaustive model, the service system reference model establishes a foundation upon which we model the service concept to concretize the service strategy of what to deliver and how that strategy should be designed as a combination of service system components. The service characteristics view establishes a goal-oriented approach to develop services driven by customer needs. The requirement model incites that all service stakeholders should collaborate with their skills and ingenuity to refine customer needs, wants, and demands into service requirements to design and develop innovative services. Interested readers may refer to [122, 123, 126, 127] for more information about the IT-enabled service system modeling framework, namely, the service system reference model, the requirement model, and mapping algorithms. Additional information about specifying business requirements with SBVR facts and rules in the context of services are explained in [125] and [124].

Chapter 3
Collaborative Design Methods Driven by Business Artifacts

Keywords Service collaborative design · Business artifacts · Service processes · Business processes · Design patterns · People-centric processes · Collaboration patterns · Service process model · Service life cycle · Business artifact discovery

3.1 Introduction

In the context of IT-enabled services, service processes are ad hoc activities and focus on what can be done to achieve the service strategy (i.e., service concept) and progressively decide with customers on appropriate actions to satisfy their needs [103]. Building services processes reveal several challenges. The role of service processes aims at assisting service stakeholders to collaborate and innovate rather than determining what are the subsequent activities to perform and what is their execution order [98]. Service processes are incremental processes, requiring customer participation to actively take part in the service design and delivery [105]. By such, customers' involvement is regarded as the most salient difference between production processes and service processes [103]. They also involve low level of repeatability, making them difficult to be automated and managed with business process languages [76], business process models [102], and workflows [146].

IT-enabled services are seen as information-driven, customer-centric, e-oriented, and productivity-focused [17]. In the introduction chapter, we demonstrate that services processes are also information-intensive and people-intensive activities rather than controlled routine business activities [25, 50, 90]. Without excluding their utilities in automating controlled routine activities, business processes—as activity-oriented processes—show drawbacks and limitations when they are applied to service processes for which there are many variations in customer inputs in collaborative and social-based environments [105]. There appears to be a severe gap between the promise of business processes of facilitating interactions between business activities and what they really offer for services in terms of collaboration in ad hoc environments. As a result, processes in services require a paradigm shift from activity-oriented processes driven by control flows to information-oriented processes driven by dataflows.

Y. Badr, *Smart Digital Service Ecosystems*, SpringerBriefs in Service Science, https://doi.org/10.1007/978-3-031-27926-3_3

In the following sections, we particularly focus on business artifacts to develop information-driven service processes and design patterns to design people-centric processes. Business artifacts overcome drawbacks of activity-oriented processes driven by predefined control flows with information-oriented processes driven by dataflows. Complementarily, design patterns deal with general solutions, including people, business, and technological aspects for recurrent problems in service design and provision in accordance with the systemic approach. Finally, we briefly present a service process as a collaboration process, including interaction patterns and collaboration patterns, the collaborative design method, and the business artifact discovery method.

3.2   Challenges Related to Service Processes

The customers participations in service processes and the variability of their inputs imply important structural changes, such as customized services or standardized services, making service processes ad hoc and unstructured processes and including divergent actions, for which the control flow cannot be defined [104]. A close look at the IHIP service characteristics shows us that services are simultaneously produced and consumed. The production and consumption are thus interactive and inseparable processes, which require simultaneous involvements of service providers and customers. This inseparability characteristic makes the service as a whole complex process of interactions between participants and includes technology, facilities, equipment, layout, and processes that generate the service outcome. Therefore, the service design process and service delivery processes are inseparable, as two faces for the same coin, and thus require a new approach to design and build them. Another important aspect in service processes is their link to collaboration; by taking a holistic point of view, we consider that service processes result from collaboration. Collaboration is the fundamental activity in services, resulting from interactions between providers and customers. Collaboration is a very complex process. Camarinha-Matos and Afsarmanesh [147] define collaboration as "*A process in which entities share information, resources and responsibilities to jointly plan, implement, and evaluate a program of activities to achieve a common goal, which implies the sharing of risks, resources, responsibilities, and mutual trust and engagement of participants to cocreate value.*" In this definition we observe the system elements (e.g., goals, participants, entities, ...) and service elements (e.g., resources, value cocreation, ...). Moreover, when collaboration occurs, the activities, which have to be executed, can be defined or may not be defined in the beginning (i.e., ad hoc activities), and providers have a variety of strategies to decide on which activity to choose. Service provider decisions will gradually determine appropriate actions to satisfy customers' needs. The main question is how to model service processes to support collaboration, for which there are many variations in customer inputs without control flows of ad hoc activities or predefined activities.

In the literature, process control approaches that may support business collaboration are divided into three approaches [148, 149]:

1. The activity-based (i.e., process-based) approach consists of tasks and control flows, indicating predetermined sequences among activities. BPEL and XPDL are noticeably examples of activity-based processes [76].
2. The data-based (i.e., information-based) approach requires no explicit control flow and consists of defined activities or tasks to manipulate data. The execution of activities become available based on data events [150]. Active XML [151] and document-driven workflows [149] are examples of data-driven processes.
3. The conversation-based (i.e., organization-based) approach is by which individual participants perform ad hoc activities, which are not explicitly defined in advance to reach business goals. The business goal is reached as a result of a series of interactions between participants. The role activity diagram (RAD) [152] is an example of techniques to model human collaboration and how people reach their business goals through communications.

The activity-based approach is the dominant approach to design and implement business processes. The scientific literature shows the proliferation of business process languages [76], process models [102], process standards [75], and workflows [146]. Business processes support service processes by automating routine activities but fail to build effective service processes for the following reasons:

1. Business process languages and their workflow management systems are too restrictive and have problems dealing with structural changes and context evolution, as well as accommodating high-variance in customer inputs [25, 153, 154]. Unlike business processes, which are driven by control flows to determine how the work should be done, service processes focus on what can be done to achieve business goals, such as satisfying customer needs and improving the service quality. Van der Aalst et al. [98] demonstrate that recent approaches to flexible workflow management systems are still based on the control flow as the only mechanism for supporting processes.
2. The activity-based modeling does not match the way people do business [155, 156]. Activities in business processes are often defined with respect to strict business logics. To process business objects during the execution of multiple business process instances, it is quite difficult to dynamically modify activities or their execution order. Each activity is also viewed as a black box and does not show the business object statuses or what business objects are updated. Although the conversation-based approach is for human-driven processes, Beeson et al. [157] highlight situations where interactions are predetermined, which make the sequencing of activities and interactions too restrictive, and the expressiveness of the process model for a dynamic environment is very low.
3. Dufresne and Martin [158] provide a brief description of some of the more prominent process modeling methods over the last 40 years (e.g., flow charts, data flow diagrams, control flow diagrams, IDEF01, etc.). The proliferation of business process languages [76], process models [102], and process stan-

dards [75] shows that specifications of service processes with rigorous and technical notations, as proposed by software engineering or as found in business processes, produce often flow diagrams that are inappropriate for customer involvement and collaboration among stakeholders from different disciplines.

Service processes are somehow similar to the case handling or the case management [159]. The case management is "a collaborative process of assessment, planning, facilitation and advocacy for options and services to meet an individual's needs through communication and available resources" [160]. Examples of cases include business requirements in research and development, call handling in call centers, analysis and study in consulting and engineering firms, etc. Cases are managed by knowledgeable workers in collaboration with customers and their management cannot be predefined in advance. They thereby allow workers leeway in determining how to handle them. In our view, the salient difference between service processes and case management is that service processes require advanced requirement assessments and continuous feedback to improve service qualities accordingly since the customer satisfaction is the goal to be reached in service design and provision. The case management often remains a paper-based process. Rooze claims that there is no clear direction in how to manage special cases often characterized by dynamic and ad hoc collaboration processes requiring human participation [161].

Services are seen as information-intensive and people-intensive activities [25, 50, 90]. Tien and Berg also assert that IT-enabled services are information-driven and customer-centric [17]. These characteristics make service processes information-based processes and require ad hoc activities without predefined control flows.

Recently, business artifacts [155], as an emergent data-centric approach, seem to offer the right means for us to tackle service processes with data-driven flows and handle data manipulation with predefined activities. Business artifacts are "business-relevant objects that are created, evolved, and (typically) archived as they pass through a business" [162]. Introduced as the basis of data-centric approaches, business artifacts address the rigidity of control flows in activity-centric approaches with a focus on how data achieve business goals.

In the service system reference model, the business view is based on business artifacts as the fundamental construct to define business models and business rules. Artifacts are self-contained business records that include attributes, states, and life cycles that reflect the changes in these states. The business artifact does not only describe a business entity but also encompasses knowledge about what to process without explaining how to do it. Changes in business artifact attributes and states are the result of executing one or more activities that manipulate attributes and states in accordance to their life cycles. A business artifact is information record used by a business managers to manage their businesses [163]. The value of an artifact relies on its data model that is primarily manageable, analyzable, and flexible from the perspective of business managers. Compared to business processes, business artifacts are mainly data-centric processes, whereby control flows are based on business state transitions and business rules [155, 160]. Through the use of business artifacts, it is possible to make service processes focus on business goals rather than defining activities and automating their executions in business processes. Business professionals thus seamlessly manage activities and intuitively construct collaborations with partners by exchanging business artifacts.

To adopt the collaboration among service stakeholders as a basis for designing IT-enabled services, business artifact-driven processes should be developed and integrated into IT-enabled service design method and service processes. Building service processes with business artifacts, thus, makes possible to release a shift from activity-driven processes to information-driven processes. However, the application of business artifacts is not straightforward. Many unanswered questions remain about business artifacts in the context of IT-enabled services.

3.3   Service Collaborative Design and Processes with Business Artifacts

After all, service processes are people processes: Customers do not only express their needs and requirements, but they also take part in all levels of decisions regarding service design, delivery, and consumption. Service providers, as knowledgeable workers, have to continuously assess customer needs and inputs, after which further actions have to be scheduled accordingly. Service processes are also information-driven processes: Rather than being predefined and deterministic processes, they continuously evolve during service production and consumption in response to exchanged information among service stakeholders to achieve common goals.

To specify and build collaboration processes, we developed several contributions to build service processes as information-intensive and people-centric processes, which cannot be handled by activity-based processes (i.e., business processes).

To this end, we proposed a data-driven approach to build information-driven processes [164] and design patterns to enable a problem-solving mindset and build collaboration patterns [165]. We also developed ad hoc and dynamic collaboration approach enabled by Web services [166] and a novel method to discover and build business artifact from business data and operations [167]. In the following sections, we briefly present these contributions within the context of architectural framework for IT-enabled services. In particular, we present the following contributions:

1. The **collaboration model**: hinges on a top-level ontology to define major concepts related to collaboration, such as dataflow, business goals, actors, and exchanged information. The collaboration model includes:

    (a) The *interaction patterns* provide different scenarios to coordinate collaboration and possible interactions by exchanging business artifacts.
    (b) The *service process model* relies on dataflows to exchange business artifacts based on the interaction patterns and the collaboration model.

2. The **collaboration patterns**: They define collaboration life cycle based on five phases to describe collaboration processes as incremental and iterative processes of. Service stakeholders and customers interact in each phase following collaboration patterns, each of which relies on problem-solving templates to define the best way to collaborate by exploiting required resources, technology, service participant roles, business artifacts and rules, and delivery channels.
3. The **collaborative design method**: includes eight steps and describes how the service system reference model, the requirement model, and the collaboration model are used together to support the IT-enabled service system codesign in an incremental and progressive collaboration.
4. The **business artifact discovery method**:This method discovery and create business artifacts and their data models, states, and life cycles based on dependency relationships among data and operations that manipulate them.

The main value of the collaboration model relies on business artifact flows to enable a knowledge-driven collaborative environment, by which service providers and consumers exchange business knowledge without dealing with technical, semantic, or business interoperability or integration problems caused by business processes. Compared to business processes, information-driven service processes do not require predefined control flows between activities that process exchanged business information. Based on data-driven service processes, service stakeholders seamlessly manage their activities and intuitively construct sustainable collaborations to design and deliver IT-enabled services.

3.3.1   The Collaboration Model

The collaboration model driven by business artifacts, business rules, and dataflows is illustrated in Fig. 3.1. It is based on an ontological representation to provide a common understanding of concepts, terminologies, and vocabularies related to service participant disciplines and facilitate service actors' interactions and sharing. Collaboration is governed by business goals, which are achieved by meeting business rules expressed in the service requirement model (e.g., SBVR-based customer requirements). Service processes are driven by business knowledge and defined by selecting collaboration patterns and establishing dataflows with delivery channels to send and/or receive business artifacts and process them based on business rules.

In the context of collaboration, service consumers send, for example, business information encapsulated by business artifacts, including their states and their life cycle to service providers. Upon reception, service providers receive business artifacts, and they decide, based on current states and business rules, which actions, operations, or activities should be performed to process business information and how states should be updated with regard to artifact life cycles. Service consumers may send back updated business artifacts to service providers. Each service actor

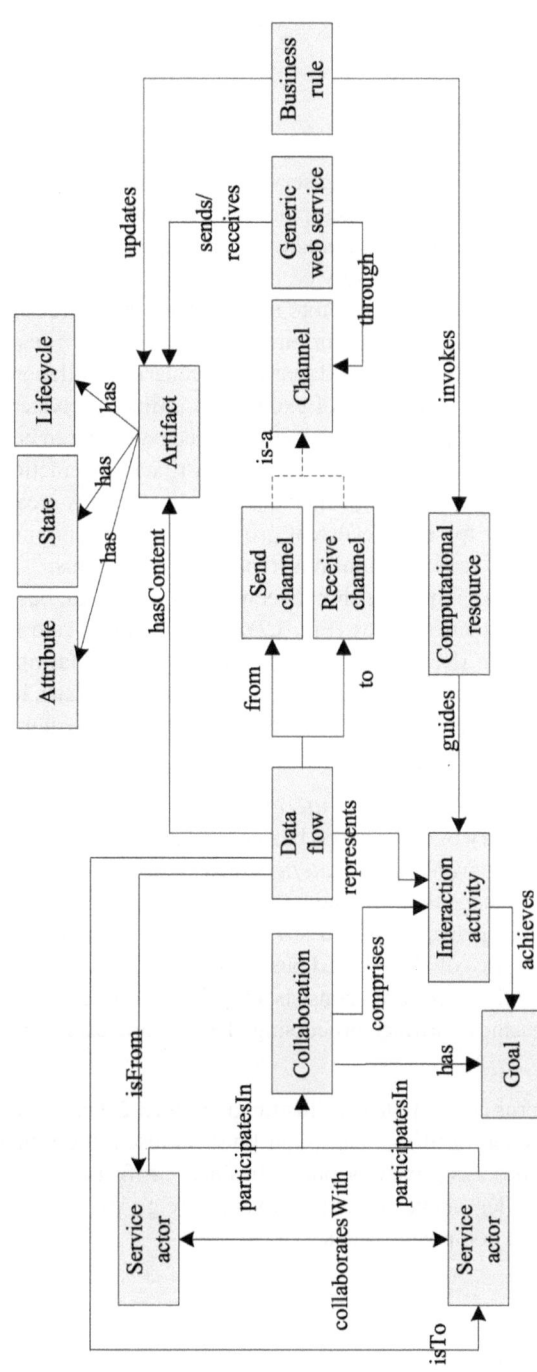

Fig. 3.1 The top-level ontology of the collaboration model

is responsible of choosing appropriate actions, such as assigning business artifact processing to knowledgeable workers or invoking Web services.

3.3.1.1 Interaction Patterns

Different collaboration scenarios among service stakeholders can be identified. For examples, a simple collaboration takes place by sending business artifacts from one service actor to another (i.e., unidirectional scenario), whereas a bidirectional collaboration may include several service actors who send and receive business artifacts. In other situations, service actors are involved in several disjoint collaborations and concurrently exchange various business artifacts (i.e., coordination scenario). In order to capture these collaboration scenarios, collaboration interactions can be specified with dataflows of exchanged artifacts between senders and receivers as interaction patterns. A collaboration-based patterns can be used to build complex collaboration scenarios driven by business artifacts. Interaction patterns are inspired by the design patterns [168], which are common in many areas of computer science to describe how to solve recurrent problems in different situations [169]. Patterns are thus general reusable solutions to occurring problems.

The interaction patterns distinguish between public or private business artifacts and shared or private business rules (Fig. 3.2). Each business rule has an Event-if-Condition-then-Action (ECA) format to trigger an action that manipulates the business artifact data model (i.e., attribute-value pairs) and updates its states when the event occurs, and the condition holds. The interaction patterns are briefly described as follows:

- *Public artifacts and shared rules pattern*: consists of common public artifacts and a shared repository of ECA-based business rules
- *Shared events and private rules pattern*: consists of event exchange. Service actors separately manage their private business artifacts.
- *Public artifacts and private rules pattern*: consists of common public artifacts and private repositories of ECA-based business rules.
- *Subcontracted artifacts pattern*: consists of private artifacts and subcontracts some parts of business artifact processing. Events are exchanged to manage artifacts

The interaction patterns are graphically illustrated in Fig. 3.2. Dashed lines represent the exchange of business artifacts, and plain lines represent the transitions within their life cycles. To process private or public business artifacts with business rules, actions triggered by rules can be defined in advanced (e.g., Web services) or created on the fly (e.g., ad hoc composite Web services).

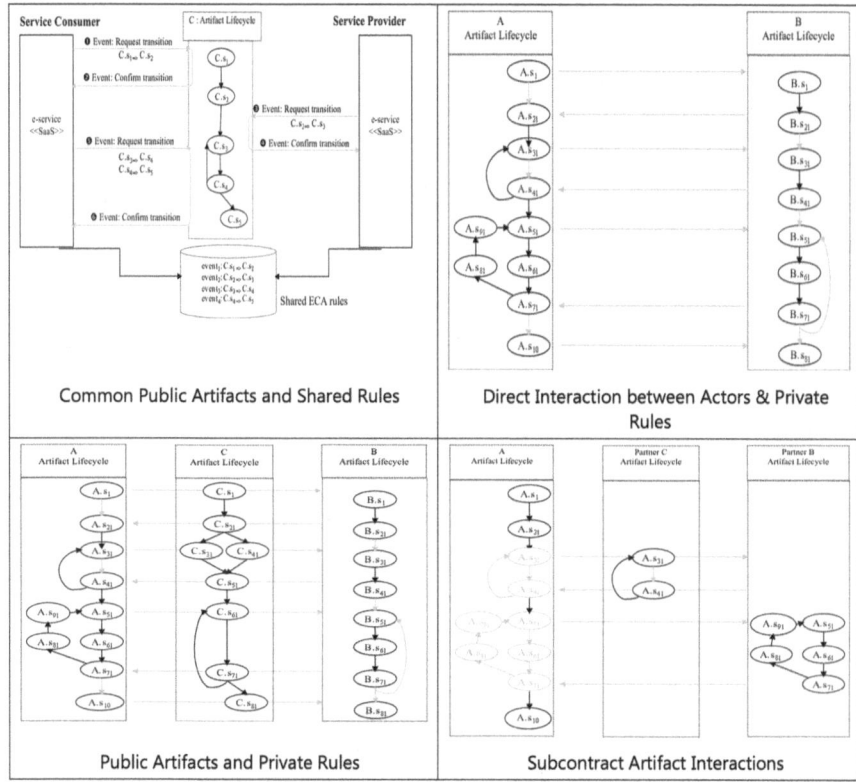

Fig. 3.2 Artifact interaction patterns

3.3.1.2   The Service Process Model

Service processes are modeled as dataflows to exchange business artifacts between
service actors involved in collaboration scenarios. Each service process is described
in terms of concepts from the top-level collaboration ontology such as interaction
patterns and dataflows, business artifacts, business rules, and delivery channels (see
Fig. 3.1). Service participants rely on their generalized ontologies and/or specialized
ontologies to refine the service processes by only specifying dataflows with public
artifacts to exchange with partners. Since private artifacts are not exchanged, service
partner can preserve their privacy when they manipulate their private artifacts. In
this context, mapping rules should be defined to support conversions between public
artifacts (Artifact$^+$) to private artifacts (Artifact$^-$) and vice versa. The collaboration
scenario among service actors is formally defined with types of delivery channels to
exchange business artifacts as follows:

Channel = { Schannel ∪ RChannel }, where

SChannel = { Artifact$^-$(?P) ∧ Rn(?P, ?B) => Artifact$^+$(?B) } is the receiver
channel

RChannel = { Artifact$^+$(?B)$\wedge$ Re(?B, ?P) => Artifact$^-$(?P) } is the sender channel.

The delivery channel allows service actors (i.e., senders/receivers) to establish dataflows. Each service actor ensures the mapping from private artifacts and public artifacts by the rendering function (Rn) or between the public artifact and the private artifact by the reverse function (Re).

The mapping rules are used to implement Re and Rn functions to translate private artifacts to public artifacts and vice versa.

MappingRules = { ReverseRules $\cup$ RenderRules }

ReverseRules = {Re(?B, ?P) / Artifact$^+$(?B) $\wedge$ Artifact$^-$(?P) $\wedge$ Attribute(?B, b) $\wedge$ State(?B, s) $\wedge$ ... => Attribute(?P, b) $\wedge$State(?P, s) $\wedge$ ... }

RenderRules = {Rn(?P, ?B) / Artifact$^+$(?B) $\wedge$ Artifact$^-$(?P) $\wedge$ Attribute(?P, b) $\wedge$ State(?P, s) $\wedge$ ... => Attribute(?B, b) $\wedge$ State(?B, s) $\wedge$ ... }

Finally, the collaboration is defined as a set of service processes. Each service process is defined as a tuple of a dataflow, a set of business artifacts, and an interaction pattern:

Service Process={ <Flow, Artifacts, interaction pattern> }

where Artifact={Artifact$^+\cup$ Artifact$^-$}

Each flow is defined by sender and receiver channels.

Flow = {(SC, B, RC) / SChannel(?SC) $\wedge$ Artifact$^+$(?B) => RChannel(?RC)}

In order to manipulate business artifacts, we define business rules based on the Event-if-Condition-then-Action (ECA) format to associate business artifacts with appropriate actions, which could be manual tasks, software applications or Web services, or composite Web services (i.e., business processes). Based on the Semantic Web Rule Language (SWRL), we define a condition, c_i, as a conjunctive logical expression over predicates, representing artifact attributes and states as follows:

c_i = Attribute(?A, ?a) $\wedge$ State(?A, ?s) $\wedge$ ...

The set of business rules that manipulate business artifact according to their life cycles is defined as:

BusinessRules = { Event: **if** ($c_{PRE}\wedge c_I$) **then Invoke** Action$_i$ **and** ($c_O\wedge c_{EFF}$) }

where c_{PRE}, c_I, c_O, and c_{EFF} are, respectively, conditions over PRE, I, O ,and EFF.

The PRE is the set of preconditions, the I is the set of inputs, the O is the set of output, and the EFF is the set of effects (see Figs. 3.2, 3.3, 3.4, and 3.5). Depending on current artifact attribute values and states in the preconditions (PRE) and the inputs set (I), the business rule invokes an action/task to update artifact attributes and states in the outputs (O) and effects (PRE) sets.

The main value of the service process model driven by business artifacts is its fine-grained structure, namely, mapping rules, flows (e.g., pair of channels), service processes (e.g., flow-artifacts-pattern tuples), and business rules. This structure allows a great flexibility to manage or customized each aspect of the service process and, consequently, the collaboration

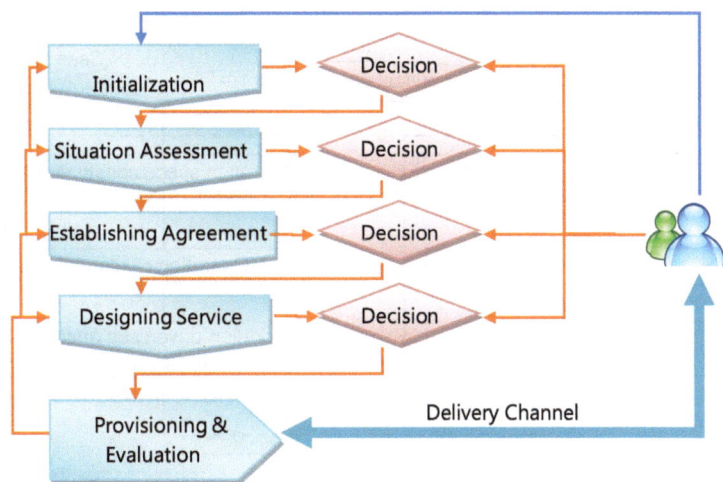

Fig. 3.3 Collaboration process life cycle

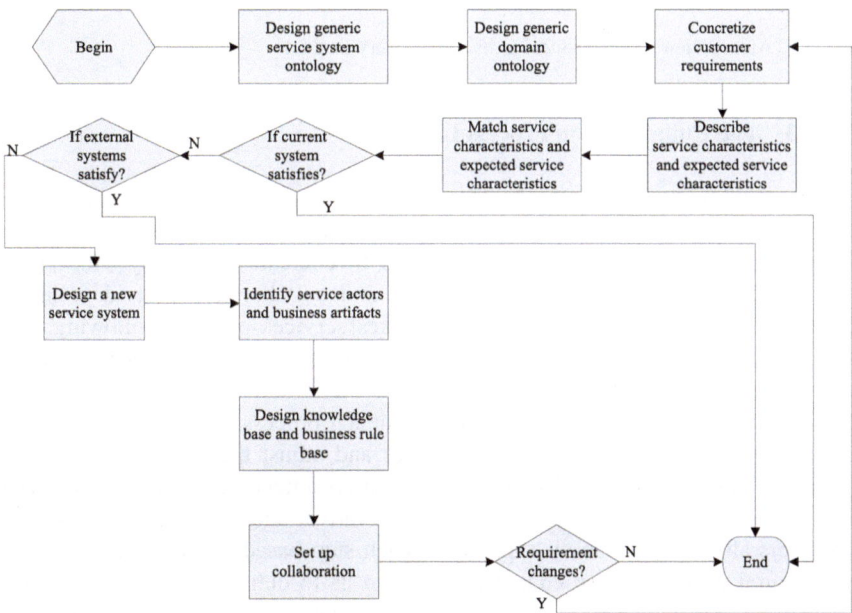

Fig. 3.4 The flowchart of the collaborative design method

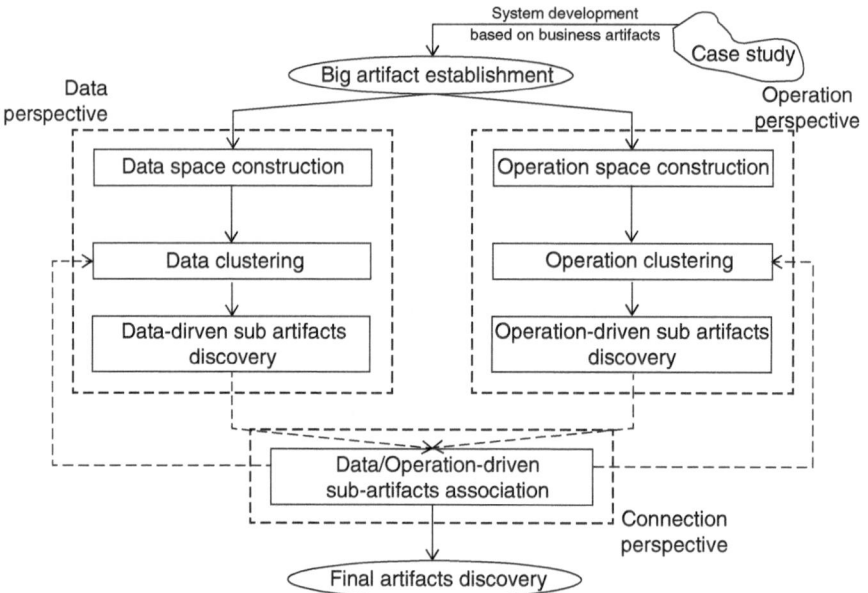

Fig. 3.5 An overview of the business artifacts discovery method

3.3.1.3 The Collaboration Process Life Cycle

The service life cycle refers to the succession of phases over time covering service design, delivery, and consumption from its creation to completion. Effectively managing the service life cycle is essential to cocreate service systems [170]. The correlation between service life cycle and collaboration is evident in IT-enabled services, because collaboration makes services progress following their service life cycle phases. Compared to traditional software development processes or manufacturing processes, the service life cycle is not straightforward or linear process. It designates progressive and incremental process with possibilities to go back to previous phases to manage changes and adjust the service performance based on customer feedbacks. In fact, customers intervene at each step during service design and consumption (i.e., collaborative, and cyclic design). Service providers also improve service quality at each step based on consumer feedback (e.g., spiral design). These characteristics lead us to define the collaboration life cycle in Fig. 3.3, which includes five phases to ensure progressive, incremental, and agile implementation of IT-enabled services, which we briefly introduce them as follows:

Initialization phase: entails customer requests to enable traditional services with ICTs to improve their service qualities, satisfy customer needs, improve customer experiences, add value to their service through innovative solutions, or create new business models with capabilities offered by ICT-enabled environments. It includes activities such as establish contacts with customers and gather information about their services.

Situation assessment phase: conducts preliminary studies to analyze customer requests and identify business goals and opportunities. It comprises activities such as identify customer needs, investigate innovation opportunities, and establish a service strategy and identify service characteristics.

Establishing an agreement phase: sets up business contracts and service level agreements between service providers and consumers. It includes activities such as assess consumers constraints (e.g., budget, milestones,...) and service expected characteristics and features.

Designing service phase: involves the organization of service provider's resources, competencies, skills, infrastructures, technologies, time, and effort to design and deliver customized service systems. It consists of activities such as requirement assessment, business artifact discovery, collaborative design, and service bundling.

Provisioning service and evaluation phase: consists of exchanging business artifacts and mobilizing an appropriate level of resources, ingenuity, and skills to produce benefits with service consumers. It relies on activities such as perform operations, adapt services to changes, ensue service quality, and improve customer satisfaction.

Following the collaboration life cycle, it is worth to mention that customer intervenes at each phase and decides with the service provider whether they move ahead to the following phase, review the current phase, or undergo any backward phases, unless customer satisfaction is fulfilled. In addition, each phase is described by a set of collaboration patterns as explained in the following paragraphs to facilitate collaboration.

3.3.1.4   Collaboration Patterns

When designing IT-enabled services, service providers and customers could efficiently collaborate and innovate with efficient solutions to solve problems, if they refer to generic patterns, integrating business, technology, and participant competencies. Patterns are general reusable solutions to commonly occurring problems. Each pattern is a template of recurrent interactions. Gottesdiener [171] classifies patterns as technical patterns, structural patterns, behavioral patterns, and concurrency patterns. In our data-driven collaboration approach, collaboration can be understood in terms of what service actors can do when they exchange business artifacts [172]. To this end, we define a set of collaboration patterns to be used in each phase in the collaboration life cycle (Fig. 3.3). In the broad sense, collaboration patterns are arrangement of social (people, roles, social networks,...), business (rules, markets, benefits, ...), and technical (smart phones, sensors, ...) issues. They capture best practices about recurring collaborative problems and solutions; facilitate informal, dynamic, and ad hoc information-intensive collaboration; capture knowledge exchanges; and trigger human or machine operations in response to particular events in a specific context [172].

As illustrated in Table 3.1, a typical collaboration pattern is composed of a goal describing its purpose, a generic problem description, information describing its context, a solution respectively consisting of business artifacts describing exchanged knowledge, interaction patterns, delivery channels, and service processes triggered by business rules within a specific context and in response to events.

The main value of the collaboration life cycle is its abilities to simultaneously integrate design and delivery processes with backward paths to any phase. However, the collaboration life cycle is not covered exhaustively in our research activities notwithstanding the completeness of our work and contributions to the collaboration life cycle. For example, the first three phases provide only general guidelines found in project management processes.

3.3.1.5 The Collaborative Service Design Method

By focusing on the service design phase and the provision/evaluation phase, we develop the collaborative service design method for IT-enabled services based on the prerequisites of the service design model, namely, service concept, service processes, and the service system [47]. The collaborative design method is similar in spirit to agile software development process in software engineering [138], which is iterative and incremental development, where requirements and solutions evolve through collaboration between self-organized teams. Conversely, requirements in the context of IT-enabled services are not limited to software development. They mainly concern business and socio-technical aspects and engage stakeholders from different disciplines. The design method refers to a set of steps to guide the design and implementation of IT-enabled service systems based on the service system reference model, the requirement model, and the collaboration model (see Fig. 3.4). The collaborative design method includes eight steps:

Step 1: Designing Generic Service System Ontologies Service providers at first identify fundamental service system elements that build up their businesses and service strategy based on the systems view (i.e., competencies, resources, processes, technology, etc.). They then create the top-level service system ontology to describe the general service concepts that are independent from any particular domain.

Step 2: Designing Domain Ontologies In response to a request to enable a particular service with ICT, service providers refine their generalized ontologies, while they collaborate with the customer and service stakeholders. Service providers design and build specialized ontologies to describe domain specific concepts related to the customer services and businesses.

Step 3: Concretizing Customer Needs into Requirements Based on the collaboration model, service providers and customers at first collaborate with each other to concretize customer needs. At this step, needs are specified with the graphical i*-

Table 3.1 Collaboration pattern general template structure

Field	Description	Example
Pattern name	Collaboration pattern's name	Confirm a claim
Goal and problem	A description of the problem and hard and soft goals to achieve (cf. SRML)	Evaluate eligibility of customer claims
Context	Declare the domain where the collaboration is applicable	Online insurance service
Referred artifacts	Refer to exchanged artifacts in this pattern	Customer, order, claim, refund
Referred roles	Service actors that take part in this pattern	Customers, salesmen, accounts
Preconditions	Refer to conditions that define the context and before applying the pattern	Order.Paid = True
Events	Trigger business rules evaluation	OrderItem.Quality = Low
Exception	Handle unexpected deviations from an ideal sequence of collaboration caused by resource failures, requirement changes...)	Assign evaluation to external experts
Post-conditions	List effects that must be satisfied after applying the pattern	Claim.Confirmed = True Create the Refund Artifact
Interaction patterns	Identify artifact interaction patterns	– Shared business artifacts: claim—list of private business rules
Delivery channels	Choose delivery channels to exchange business artifacts	– Channel for the claim artifact: email ∨ postal mail ∨ phone
Related patterns	(1) Collaboration patterns that optionally can be executed in parallel or after termination	(1) Subsequent collaboration pattern: create refund
	(2) Alternative collaboration patterns that can be used instead of the described one	(2) Alternative collaboration pattern: none
	(3) Conflicting collaboration patterns that cannot be executed concurrently	(3) Conflicting collaboration patterns: refuse claim
Service processes	Describe how this pattern deals with the problem expressed with soft and hard goals.—List of services processes (artifacts, channels, mapping rules, transition rules)	

goal oriented model and refined into wants expressed with a set of characteristics, describing the expected service. Each characteristic is composed of features and constraints. Business experts and IT specialists reformulate the customer needs with the Service Requirement Modeling Language (SRML) to specify expected service characteristics, by representing their quantitative and qualitative features and ensure QoS. Customer requirements are formally expressed with SBVR concepts, facilitating the generation of required business artifacts and SWRL rules.

Step 4: Customizing Existing Services and/or Adopting Third Party Services
Based on customer requirements, service providers assess whether exposed service characteristics of existing IT-enabled services satisfy customers by matching expected service characteristics. Service providers may outsource, inquiry for third party services or adopt an incremental innovation approach, consisting of modifying existing service systems by collaboration with customers to create tailored solutions for customer problems. If there is no match, a complete new service system has to be created. IT specialists identify required service system components with goal-oriented graphical model (soft/hard goals).

Step 5: Identifying Exchanged Knowledge In IT-enabled service systems, service actors exchange explicit business knowledge during their collaboration by virtue of business artifacts. Service designers discover business artifacts. In the next section, we present a business artifact discovery method.

Step 6: Identifying Business Rules IT specialists collaborate with business experts to enrich the generic and specialized domain ontologies. Based on customer requirements, they define business rules from in natural-structured formal language (i.e., SBVR), and they thus map them to SWRL rules. Each SWRL rule invokes software applications. This rule-based approach has the advantage of excluding the business logic from software applications. The set of SWRL rules are easy to manipulate and update when requirements change without modifying software applications.

Step 7: Setting Up Collaboration Processes Service providers rely on their collaboration with customers to decide on actions to satisfy customer requirements and identify business artifacts and business rules. The collaboration also covers service design and consumption alongside with customers. Since artifacts are self-contained data entities, service participants select appropriate computational resources (e.g., Web services or SaaS) to update artifact attributes and current states with respect to their life cycles. Building data-driven service processes based on rules and exchanged business artifacts have the advantage of choosing collaboration patterns that improve the whole service.

Step 8: Requirement Changes Since collaboration is incremental and iterative process, any changes to initial requirements or any new requirements will be captured and propagated as described by the mapping algorithms in the requirements model.

3.3.2   The Business Artifact Discovery Method

Due to the importance of business artifacts in our research strategy to design framework, identifying business artifacts becomes an essential task in designing and implementing IT-enabled services and remains an open research question despite several initiatives [173]. There is an abundant literature on business artifacts [163, 174]; However, there is still a lack of formal method that assist those in charge of discovering business artifacts and making them available to situation such as service systems. The importance of such approaches is stressed in the Bhattacharya's work who developed a top-down data-centric design method for business processes [175]. Although the first step in this method is to discover business artifacts, the techniques in this step remain "subjective" by relying on the analytical skills and experiences of system analysts. Quoted from Bhattacharya et al.'s paper, "*[. . .] identifying artifacts require an understanding of the whole business process, how data are changed and shared through the process, and what data hold critical business process information. This is done through a combination of top-down analysis and by examining typical scenarios (normal business cases and exceptional cases*". Unfortunately, questions like how to discover artifacts, how artifacts can be associated with business requirements, what is the granularity level of artifacts, how artifacts are connected to each other, and how data or operations support the rationale of artifacts are not well addressed in the business artifact community.

Acknowledging the complexity of the business artifact discovery, we develop a method to discover business artifacts. To the best of our knowledge, this is the first work of its kind to present a complete method for the business artifact discovery.

The business artifacts discovery method is illustrated in Fig. 3.5 and relies on three perspectives to discover business artifact attributes, life cycles, and related operations (tasks) that manipulate them. Unlike the Bhattacharya's top-down approach, the business artifacts discovery approach is bottom-up and seeks to gather elementary data and establish dependency relationships among them to define the artifact granularity. In the following paragraphs, the business artifact discovery method is briefly presented from the data perspective, operation perspective, and connection perspective.

- The *data perspective* identifies data that are manipulated and exchanged in service systems and the dependencies between these data.
- The *operation perspective* identifies operations that are executed data and the dependencies between these operations.
- The *connection perspective* establishes links between data and operations perspectives so that a list of final artifacts can be discovered.

The data and operation perspectives rely on the scope of the case study, which represents the domain to analyze in order to discover business artifacts. The discovery strategy consists of putting data and operations together to constitute the basis for establishing the *Big Artifact* (BA). The clustering algorithms creates intermediary *sub-artifacts* (SA), which are refined through a set of dependency relations in order to create the set of *final artifacts* (FA).

Firstly, the big artifact (BA) as a melting pot that includes all data and operations without any distinction to their types, natures, restrictions, roles, etc.

- *BA = <DATA, OPERATION >* where:
- *DATA* is a finite set of data d_i, $i=1.. n$,
- *OPERATION* is a finite set of operations op_j, $j=1..m$.

In a manner consistent with [156], the business artifact is defined as a container of data items, describing a business entity. Formally, a business artifact, A, is recursively defined as $A = \{d_i\}$, where d_i is either a data variable or itself an artifact. Data variables can be ordered, unordered, simple objects, complex objects, or finite or infinite data through a recursive definition. The data models in artifacts can be represented as nested name-value pairs, entity relationship schemas, or might be based on a description logic [176]. We define the artifact life cycle of *artifact A*, as $ALC_A = (S, s_0, S_f, L, T)$, where S is the set of states, s_0 is the initial state, S_f is the set of final states, L is a set of transition labels, and T is a set of transition rules. A label $l \in L$ consists of three components (event E, condition C, and action A). A transition rule is defined from a state to another as $t = (s_i, l, s_j)$, where s_i and $s_j \in ALC$.

3.3.2.1 The Data Perspective

Establishing the data perspective consists of three steps:

1. Data space construction is based on the algorithm in (Fig. 3.6 right side) to extract all data d_i of the case study the *BA* and prune their names from synonyms, homonyms, and antinomes. The algorithm also annotates each $d_i \in DATA$ with a meta-description structure as follows:

$< id, label, type_{[atomic \,|composite]}, input_{[assigned \,|calculated]},$
$use_{[update \,|consultation, \|both]}, restriction_{[read \, only \,|notransfer]} >$

2. Data clustering within this data space: Data is grouped into clusters. Each cluster is a potential candidate to map into a data-driven SA. For data clustering, we define three types of dependencies between data as follows:

- *Update dependency (UD)*: $\exists\ UD(d_i, d_j)$, if the successful update of d_i triggers the update of d_j
- *Substitution dependency (SD)*: $\exists\ SD(d_i, d_j)$, if the unavailability of d_i makes d_j available for use (due to some restrictions)
- *Removal dependency (RD)*: $\exists\ RD\ (d_i, d_j)$, if the deletion of d_i is subject to the successful deletion of d_j (RDs can be regarded as similar to data integrity constraints in databases)

In addition, each type of dependency is classified into strong and weak dependencies. The weak type is motivated by the restrictions that can be put on data. Moreover, the heuristic rules (Fig. 3.6 left side) are defined to discover dependencies

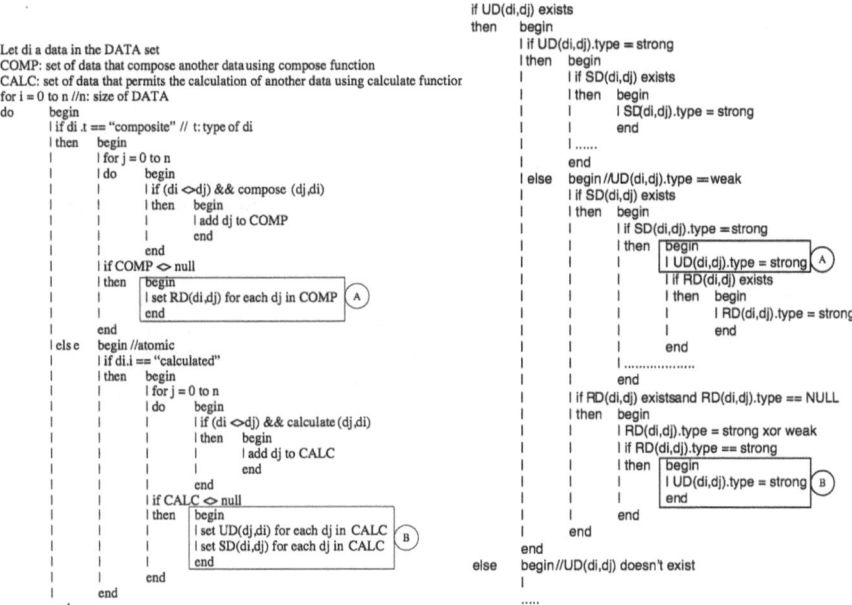

Fig. 3.6 The data dependencies algorithm and the consistency checking algorithm

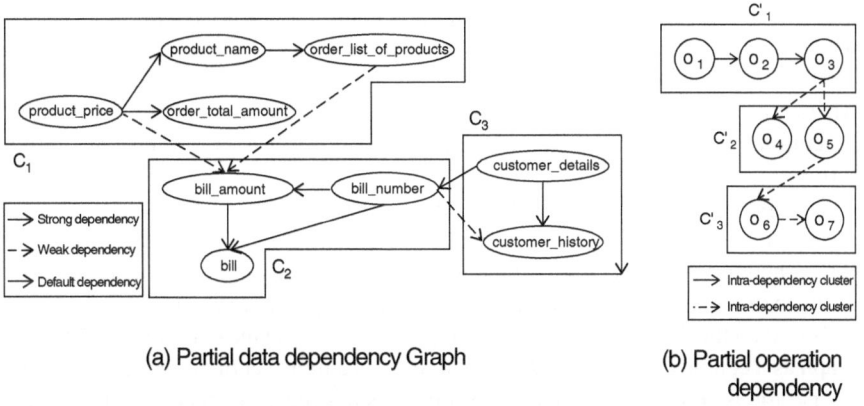

(a) Partial data dependency Graph

(b) Partial operation dependency

Fig. 3.7 The data dependency graph

between data and check their consistency. These rules make possible to modify their respective types suitably from strong to weak and vice versa, so as to ensure that dependencies are consistent and deadlock free.

3. Discovery of data-driven sub-artifacts: Based on the dependencies between data, we build the data dependency graph as illustrated in Fig. 3.7a. The data dependency graph is a couple $G_d = <N_d, E_d>$ where:

N_d is a finite set of nodes that correspond to data obtained after completing the first step of the data space construction

$E_d = (d_i, d_j, Att)$ is a set of edges that connect d_i to d_j, and Att defines whether (i) the dependency is a default one, or (ii) if not, whether the dependency is strong or weak.

Figure 3.7a shows that strong dependencies with plain lines and weak dependencies with dashed lines. We assume the existence of "default" strong dependencies between atomic data that belong to the same composite data, and this is done to preserve data consistency.

Figure 3.7b shows that if the weak edges in the dependency graph are omitted, the resultant graph is disconnected, partitioned into one or more connected *Clusters* (C_i). Each cluster is a candidate to become a data-driven SA. The connected components can be determined in linear time based on the number of nodes of the graph [177].

3.3.2.2 The Operation Perspective

Establishing the operation perspective consists of steps:

1. Operation space construction aims to extract all operations o_i, of the case study the *OPERATION* of the BA, and prune their names from synonyms and homonyms. The algorithm annotates each o_i, $\in$ *OPERATION* with a meta-description structure denoted by the tuple: *<input, output, preconditions, effects>* where input and output are set of data reported in *DATA* of the *BA*, preconditions is a set of Boolean formulae that must hold true if operation o_i is to start execution, whereas effects is a set of Boolean formulae that hold true after operation o_i finishes execution. These formulae are assumed to be described in conjunctive normal forms.

2. Operation clustering within the operation space: We say that o_j is dependent on o_i if the following condition holds: One of the outputs of o_i is an input for o_j, and if $e_i \in$ *effects* is an effect of o_i and $p_j \in$ *preconditions* is a precondition of o_j, then $e_i \Rightarrow p_j$; i.e., the effect e_i subsumes the *precondition* p_j.

Based on the dependency between outputs, we develop the operation dependency graph as a couple $G_o = <N_o, E_o>$, where N_o is a finite set of nodes corresponding to operations and $E_o = (o_i, o_j)$ is a set of directed edges connecting o_i to o_j, such that o_j is dependent on o_i. The only assumption that is made about the operation dependency graph is that it should be acyclic.

3. Operation-Driven Sub-Artifacts Discovery: Based on the acyclic operation dependency graph, we developed the establishing operation dependency graph algorithm to group the operations that use the same set of input data into clusters and then relate these clusters to potential *operation-driven SA*. Each cluster derived using the operation clustering algorithm is a potential operation-driven SA. The reason is to link the operations that manipulate a common set of data into a common container.

Figure 3.7b depicts the operation dependency graph. Plain lines represent intra-dependencies between operations in the same clusters, while the dashed lines represent interdependencies between operations in separate clusters.

3.3.2.3 The Common Perspective

In this perspective, the *data-driven* SAs and *operation-driven* SAs obtained previously are combined to form the future data-driven final artifacts. The idea of the connection perspective is that the *data-driven* SAs provide the necessary data and the *operation-driven SAs* provide the necessary operations that act on these data.

The *final artifact* (*FA*) is defined in a business case study as:

$FA = <d\text{-}SA_i, O\text{-}SA, STATE, ALC>$ where:

- $d\text{-}SA_i$ is a specific *data-driven SA* *reported in the data* perspective. O-SA is a finite state of all the operation-driven SAs reported in the operation perspective that act on the d-SAi in terms of input and output actions.
- *STATE* is a finite set of all the states that constitute the life cycle of *FA*.
- A state *s* consists of a set of data extracted from the $d\text{-}SA_i$.
- *ALC i*s the life cycle of the *FA* that is built upon the states of *STATE*.

Based on the example depicted in Fig. 3.8, the generation of the *final artifact FA$_j$* can be illustrated with the following steps:

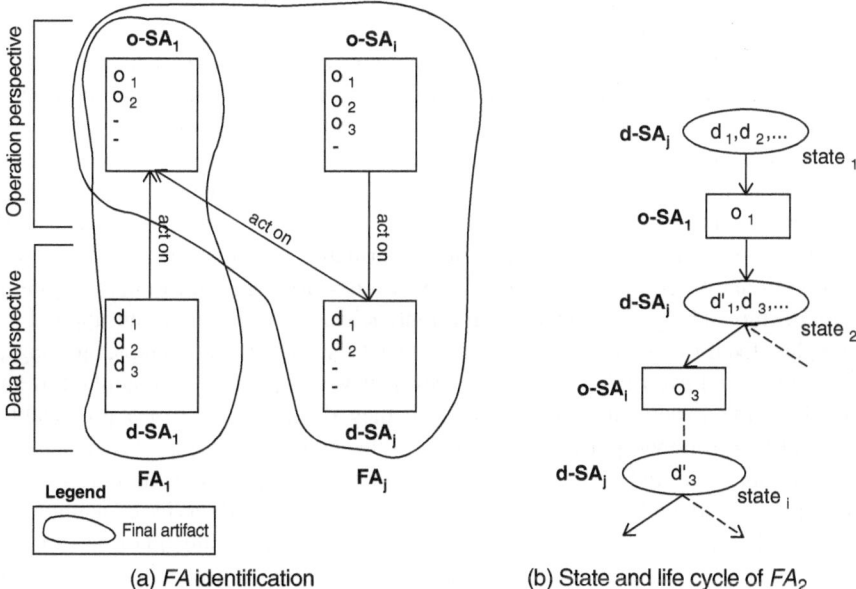

(a) *FA* identification   (b) State and life cycle of FA_2

Fig. 3.8 Connecting data and operation perspectives

- $d - SA_j = \{d_1, d_2\}$.
- $O - SA = \{o\text{-}SA1, o\text{-}SA_i\}$ where $o\text{-}SA_1 = \{o_1, o_2\}$ and $o\text{-}SA_i = \{o_1, o_2, o_3\}$.
- $STATE = \{state_1, state_2, \ldots, state_i, \ldots\}$where

  - $state_1 = \{d_1, d_2\}$, $state_2 = \{d'_1, d_3\}$, and $state_i = \{d'_3\}$, $\ldots$

- ALC = $\{(state_1, o_1, state_2), (state_2, o_3, state_i), (state_2, o_?, state_?), \ldots \}$

3.4 Concluding Remarks

In the realm of IT-enabled services, we argue that collaboration is the most important aspect and is the challenge that defeats most of traditional approaches that build attempt to build collaborative with activity-driven processes. Activity-driven modeling does not match the way people do business [155]. Activity-based modeling techniques [75], which mainly deal with business processes, fail to address human-driven collaboration and require that business specialists, analysts, and informatics experts define in advance business processes using technical-oriented languages. In our research, we confronted the inadequacy of business processes to model interactions and collaboration in IT-enabled services. We tackle the challenge of data-driven collaboration by means of business artifacts. We also propose business artifact discovery method and a collaborative design method.

As for the collaborative design method, it may look similar to traditional methods for designing information systems or in software engineering. Nevertheless, it somehow differs in many aspects, such as the context in which we apply the method. The collaborative design method takes into consideration service characteristics, the collaboration of various actors with different backgrounds, and customer satisfaction driven by requirements. The silent feature of the collaborative design method is that it also advocates a middle-out approach, because it relies on views and models that expose an appropriate level of details understandable from business managers, service architects, to developers. However, top-down or bottom-up approaches are widely adopted to architect information systems or software systems in traditional enterprise architecture (EA) frameworks (i.e., information system asrchitecture) [178]. Top-down design approaches worked well when applied to complex, fixed functions, that is, human artefacts, such as aircraft, ships, buildings, computers, and even EA software. However, it works poorly when applied to an equally wide variety of domains, because they do not behave in a predictable way [117]. Top-down approaches are not flexible enough to encompass dynamic (ad hoc) collaboration and not flexible enough to mobilize external and internal business resources and adapt them to changes during service design and service consumption. Conversely, the bottom-up design approaches start from a quite small part of the business or system components in an organization and then gradually scales it out to the whole organization (i.e., composing various systems together) [179, 180].

As for the business artifact discovery method, it differs from other data modeling approaches. Silverston et al. [181] argue that data modeling approaches can be either bottom-up or top-down. The former gathers existing data structures from reports, forms, and screen fields to build physical data models. The latter collect data from people who know the business case to create logical data models. Several data modeling approaches have been developed to identify business artifacts. The work in [174] provides a transformation algorithm for creating business artifacts from activity-centric process models. However, in contrast to our approach, the transformation algorithm requires a complete process model description, and that approach only focuses on a "domination" dependency similar to our removal dependencies. A reverse approach proposed in [182] seeks to use artifact life cycles to generate an initial business process model that is compliant with these life cycles. The work in [183] automatically constructs declarative artifact-centric workflow from goal specifications. It includes both the business-relevant data and a specification of their life cycle and shows how artifacts evolve over time as they move through the workflow.

Chapter 4
Toward Digital Service Ecosystems

Keywords Digital ecosystem · IT-enabled services · Digital service ecosystem ·
Web service discovery · Web service selection · Web service composition ·
Service bundling · Service processes · Service bundling · Business artifacts ·
Digital ecosystem · Service characteristics · Web service composition · Service
farming

4.1 Introduction

From the systemic-thinking perspective, IT-enabled services are discussed as a fundamental prerequisite for designing service systems [48–53]. An IT-enabled service
is thus a "*[. . .] system embedded in complex and adaptive combination of resources
and stakeholders*" [22]. In order to combine service system elements, two possible
composition mechanisms from the fields of software engineering (the SOA and the
Web service composition) and service engineering (the service bundling) could be
applied. Regardless their maturities, complexity, and technical and business scopes,
current composition mechanisms reveal drawbacks and limitations when we confront them to computational resources of IT-enabled services in ad hoc, social-based,
and dynamic environments; current Web service composition mechanisms work
best with well-defined requirements on (non-)functional Web service properties and
predefined composition plans to build composite services [184–187]. Consequently,
they are not suitable for building service processes, which entail incremental
development in which requirements evolve during collaboration and activities are
ad hoc without predefined composition plans. Some works dealt with multiple
constraints on nonfunctional properties and adaptive optimization approaches [188–
195]. However, they fail to work with partial or imprecise requirements (e.g.,
ambiguities in user needs), adapt to frequent changes (e.g., new decisions and
contextual variations), and handle constraints imposed by social-based and ad
hoc environments. In the Introduction Chapter, we also examine service bundling
mechanisms to integrate two or more services into a new service. Many works
attempt to extend SOAs from a software perspective to build business service
oriented approaches [78–81]. They mainly consider that the business service design

Y. Badr, *Smart Digital Service Ecosystems*, SpringerBriefs in Service Science,
https://doi.org/10.1007/978-3-031-27926-3_4

remains software design problems and assume that they are assemblages of software components: an assumption that is not adequate with the systemic thinking that advocate cause-effect relations and feedback loops. Few works tackle the problem of service bundling from a business perspective [28, 77, 86] and focus on combining resources and trade-offs between customer demands and resources. Combining resources in services urges novel composition logic to bundle IT-enabled services and a new shift from static Web service compositions in well-defined business contexts with unchanged requirements to dynamic Web service compositions in ad hoc environments to build ad hoc activities (i.e., ad hoc composite Web services) that process business artifacts in service processes.

An IT-enabled service as a business activity aims at improving service quality, increasing customer satisfaction, and adding value [31]. As a viable system [96], the IT-enabled system should adapt itself to changes and variations in order to improve service quality, handle new requirements, and support variety of strategies and decisions to determine appropriate actions to satisfy business goals. Since ICT-enabled environments are ad-hoc, collaborative, and social-based environments, they incite complex interactions among service elements and, consequently, defy current Web service composition and service bundling mechanisms to properly create on the fly composite Web services to manipulate business artifacts, ensure service adaptability, and efficiently recombine system resources to rebuild the IT-enabled service if necessary. Nevertheless, building IT-enabled services only with service bundling and Web service composition should not dismiss general system properties, such as evolution, cybernetics, and goal-directness [106]. We should not also neglect that services are complex systems [48] and their ICT-enabled environments is also complex. The complexity should not be oversimplified to not miss opportunities in building viable and sustainable IT-enabled services. Recently, some works attempt to study the complexity of service systems and interactions within and between service systems from an ecosystem perspective (i.e., ecological systems, living systems, and viable systems) [29, 96, 97, 196, 197]. By considering service systems as ecosystems, it becomes possible to generally illustrate complex interactions between services immersed in technological, economic, political, and social environments and complex interactions between service system elements [198].

Within the ICT-enabled environment context, the concept of digital ecosystems has recently emerged and studied under two perspectives: the human-driven perspective and the technology-driven perspective [34]. The human-driven perspective attempts to imitate the ecological ecosystem by focusing on relationships between partners (e.g., collaboration, mutualism,. . .) and their dynamic behaviors [199]. The technology-driven perspective considers the digital counterparts of service systems. In both perspectives, digital components and the human part are considered, but one of them is dominant at a time. In our research activities, we have focused on the technology-driven perspective to study evolution and complexity in IT-enabled service systems. We argue that applying digital ecosystems to IT-enabled services is not only appealing from computing or software perspective [200, 201] but also highlights the complexity of interactions and relationships inside and between

IT-enabled service systems. Moreover, digital ecosystems entail emergent characteristics, for example, self-organizing, scalability, sustainability, and evolution, which open new insights into studying IT-enabled services and defining suitable measurements for analytical studies.

The development of service bundling and Web service compositions within the context of ICT-enabled environments are fundamental blocks to deploy large-scale IT-enabled services. This requires new software architectures, supporting service bundles and providing customers access interfaces to IT-enabled services. Likewise, new insights in designing IT-enabled services from the digital ecosystem perspective lead us to investigate different research directions: What types of relationships may exist between service system elements? How do they evolve in digital ecosystem? And how to build ad hoc Web service composition mechanisms to support adaptability of service processes to changes (i.e., ad hoc composite Web services)?

4.2 Challenges Related to the Digital Ecosystems

The research on digital ecosystems have been started with the OPAALS, an European research Network of Excellence funded under the European Union's 6th Framework Programme, with a series of projects, performing in the ICT for enterprise networking and grouping four clusters: business networking, digital ecosystems, ambient intelligent technologies for the product life cycle, and enterprise interoperability [202]. One of the most comprehensive definitions for the digital ecosystem is cited by Fu with a focus on the ecology [203]: "*A digital ecosystem is a digital environment populated by digital species or digital components which can be software components, applications, services, knowledge, business processes and models, training modules, contractual frameworks, law, etc.*" The basic entity in a digital ecosystem is the digital entity, which could represent software components, applications, services, IT-enabled services, knowledge, business processes, training modules, or contractual frameworks, and so on. Digital entities form digital populations. The relationships between digital entities include interactions and cooperation. However, digital ecosystems refer to a holistic view from a multidisciplinary perspective [198]: "*A digital ecosystem is a self-organizing digital infrastructure aimed at creating a digital environment for networked organizations that supports the cooperation, the knowledge sharing, the development of open and adaptive technologies and evolutionary business models.*"

Digital business ecosystems [204], innovation digital ecosystem [202], Internet of Services [205], and Internet of Things [206] are few examples of digital ecosystems that have emerged recently and have greatly expanded with Web 2.0, new devices (e.g., sensors, RFID, smart phones, . . .), and advanced networks (e.g., wireless, satellite, . . .), making possible the development of IT-enabled services at a scale never before imagined. Supporting IT-enabled services through the adoption of digital ecosystems offers new business opportunities and innovative solutions to build multi-scale IT-enabled services. However, the design of highly distributed digital environments requires large-scale and distributed software architectures to support IT-enabled service provisions and delivery channels. We have to admit that most of software architectures in the field of digital ecosystems are extensions of SOAs with ecological principles, such as the pyramid structure, feedbacks, and autoregulation.

Although service systems have been explored to some extent as ecological systems, living systems, and viable systems to study interactions among service entities and emergent characteristics (e.g., sustainability), they remain quite theoretical and lack of details for practical design guidelines [29, 96, 97, 196, 197]. They also require further details regarding their implementations in practice.

Service systems in ICT-enabled environments require scalable architectures to build large-scale IT-enabled services from the digital ecosystem perspective. Unfortunately, most of digital ecosystem architectures extend SOAs without a holistic view regarding service system structures and relationships between digital entities that affect its evolution and behavior. This means that integrating IT-enabled service systems becomes restricted to the integration of SOA-based infrastructures (e.g., business process integration) and subject to business interoperability problems [59]. Nonetheless, relationships among service entities, internal interactions and contextual interactions with ICT-enabled environments should be identified to enhance service bundling and Web service composition mechanisms.

Although SOAs and their extensions are not convenient architectures for large-scale and complex IT-enabled services, they are still very useful to some extent to expose service computational resources as reusable and interoperable software components and provide agile computational infrastructure, supporting service processes.

Nevertheless, service processes are subject to frequent changes in their control flows that specify the execution order of their actions. In service processes, decisions on actions to perform are made while the service provider collaborates with customers. The execution order of actions cannot thus be defined in advanced and cannot be modeled as business processes.

In the case of activities that process business artifacts in service processes, business processes might be used to automate the execution of routines activities. The composition of Web services has to handle well-defined business processes and their activities' execution orders. Different approaches tackle the problem of Web service composition as reported and compared in several surveys; Dustdar and Schreiner [207] classify Web service compositions into static and dynamic composition strategies concerning design time and runtime, during which Web services are composed together. Alternative surveys attempt to focus on general planning techniques [208], artificial intelligence planning techniques [209], or QoS information to derive composite services with optimal QoS [186, 210]. Many Web service composition mechanisms proceed with two phase's process to construct composite services: composition planning and Web service selecting. Firstly, the composition planning phase aims at generating a composition plan based on predefined activities to achieve. Actions are usually represented as abstract

Web services. The composition plan indicates the execution order of different abstract Web services and is generated by matching their functional properties, e.g., operations and messages [211]. Secondly, the service selection phase aims at selecting atomic Web services based on nonfunctional properties of abstract services and integrating selected Web services in the composition plan to generate composite Web services [212]. The second phase is mainly driven by local constraints on quality of service (QoS) of atomic Web services and/or global constraints concerning QoS of the resulting composite Web services.

Several approaches have been developed to tackle the problem of dynamic compositions [189, 191, 194, 213–217], adaptive compositions [190, 218, 219], and compositions in dynamic service-oriented environments [220–222]. In our review, most of these approaches are QoS-aware and recompose Web services when QoS changes at the runtime. They do not provide means to express constraints on the control flow neither on potential dependencies between Web services.

Within social-based, collaborative, and dynamic environments, Web service composition should take into account contextual information, user preferences, multiple constraints on functional and nonfunctional Web services properties, implicit and explicit relationships between Web services, incomplete requirements, and dynamic composition of Web services. The lack of work on ad hoc Web service compositions arises a drastic challenge for the development of dynamic SOAs to support the adaptability of service processes in IT-enabled services.

To address these challenges, we have studied challenges in digital ecosystems [223] and their applications to business e-services in knowledge-intensive firms [224]. The digital service ecosystem leads us to study their properties and evolution, which results in contributions such as the concept of versionology in service characteristics and SaaS-based delivery models to bundle services [225, 226]. In the context of collaboration in digital service ecosystems, we developed ad hoc Web service composition mechanisms based on structural constraints [227] and business capabilities [228]. By considering the social behavior among service stakeholders and software components, we developed new Web service discovery based on social-based relationships [229, 230] to enable collaborative organizations [223] and digital ecosystems [231] [232] and composite Web services [233]. The ad hoc Web service composition and social-based Web service discovery are extended with a framework for Web service selection based on nonfunctional properties [234], quality of service [235], and user preferences [236].

In the following sections, we briefly present these contributions within the context of digital service ecosystems and Web service composition. In particular, we present the following contributions:

1. **The digital service ecosystem**: aims at providing a sustainable environment for building scalable IT-enabled services. It focuses on the technology-driven perspective to study various relationship patterns between service system components to illustrate their structures, social behaviors, and evolution. This leads us to investigate:

- **Social-based relationships between service components**: provide implicit relationships between service system components and guidance for building service system and systems of service systems. These relationships will be used later to create service bundles and enhance Web service discovery and ad hoc Web service composition with implicit social-based relationships.
- **The Evolution of Digital Service Ecosystem s**: attempts to study factors that impact software components and lead to the evolution of the digital service ecosystem.

2. **The Service bundling mechanism**: An important aspect of our digital service ecosystem is data-driven "links" by means of business artifact to bundle IT-enabled services. Similarly, bundling IT-enabled service systems relies on business artifacts. By such, interactions between service providers and service consumers depend only on the exchanged business artifacts and their current states. To this end, we have designed and developed the following architectures to support service bundling.

   - The **Business artifact exchange architecture**: aims at implementing the exchange mechanism of business artifacts between IT-enabled service systems based on generic Web services and repositories for mapping rules and transition rules.
   - The **SaaS-based architecture for service front-office interfaces**: allows service consumers to share access and interact with the same IT-enabled service and select some of its final characteristics that might interest them. This architecture provides multi-tenant backends in compliance with the Software-as-a-Service architectures.
   - The **Management of final service characteristics:** Since all service consumers do not have same needs regarding the final service characteristics, we extend the SaaS-based for the service front-office interfaces with a measurement system, called versionology, to study how users exploit selected service characteristics, how extrapolate user needs and decide to update service characteristics or recreate new services with new characteristics.

3. **The ad hoc Web service composition algorithm**: creates ad hoc composite Web services based on several rules and contextual information to build dynamic SOAs. The algorithm relies on a set of rules; structural and dependency rules, representing social-based relationships between Web services; local and global constraints; rules generated from service consumer requirements; and business rules that manipulate business artifacts.

   In sum, the digital ecosystem concept provides original insights into building services at a large scale by focusing on business artifacts exchanged between IT-enabled services instead of building interfaces, such as API, RPC, Web services, and RESTful services, to interoperate IT-enabled service architectures (e.g., ESB, peer-to-peer, etc.). In addition, the digital ecosystem provides new perspectives to study IT-enabled services as complex systems (e.g., evolution, social-based relationships). The business artifact exchange architecture, the

SaaS-based delivery model, versioning final characteristics, and the ad hoc Web service composition are firsthand implementations of IT-enabled services.

4.3 The Digital Service Ecosystem

Developing IT-enabled service systems with different "granularity" and scales is viewed as a meta-system or a system of systems, which offers more functionalities and performance than simply the sum of its constituent systems. We introduce the digital service ecosystem as a holistic system of service systems, in which services can be designed and operated and bundled to create scalable services at large scales. The digital service ecosystem concept refers to the adoption of ICT to enable a sustainable environment, including technologies, business models, business knowledge sharing, social communities, and IT-enabled services. The digital service ecosystem also aims at studying IT-enabled services in socio-technical and sustainable environments. In our research activities, we only confine the technology-driven perspective in digital service ecosystems to provide insights into scalable digital ecosystem architectures, social-based relationships between service system components, and the digital ecosystem evolution.

4.3.1 Social-Based Relationships Between Service Components

Information and communication technologies make "*[. . .] almost anything from people, object, to process, for any organization, large or small – can become digitally aware and networked* [30]." Studying implicit or explicit connections between digital entities, and particularly software components, shows how entities depend on each other at the design time when building the service digital ecosystem architecture and what are cause-effects relationships at runtime when they interact and evolve. To build an abstract view of digital entities (e.g., software components) and their relationships, we adopt the pattern designs, which are initially developed in the field of software engineering. Design patterns are programming idioms, often used to solve a particular problem and build software that are intelligible, flexible, and generic [168]. The concept of the design pattern has already been extended and applied to business systems and control systems. For example, work in [237] presents the "business enterprise" as a system of systems (SoS) by defining the composite pattern.

Figure 4.1a shows two types of relationships: aggregation and inheritance relationships. The composite pattern shows that a composite software component may recursively consist of other software components through inheritance and/or aggregation relationships. The hierarchical relationship ends when it reaches a leaf software component.

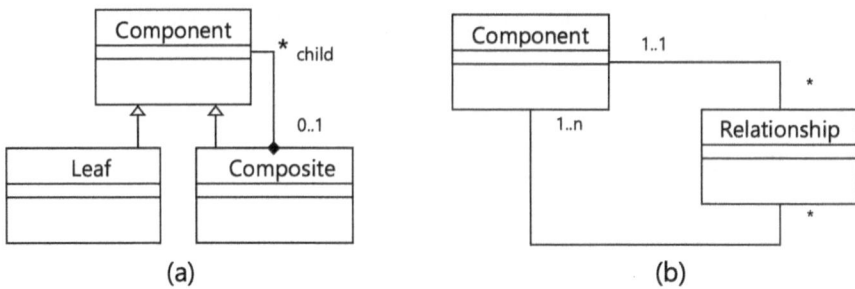

Fig. 4.1 UML composite pattern (**a**) and the relationship pattern (**b**)

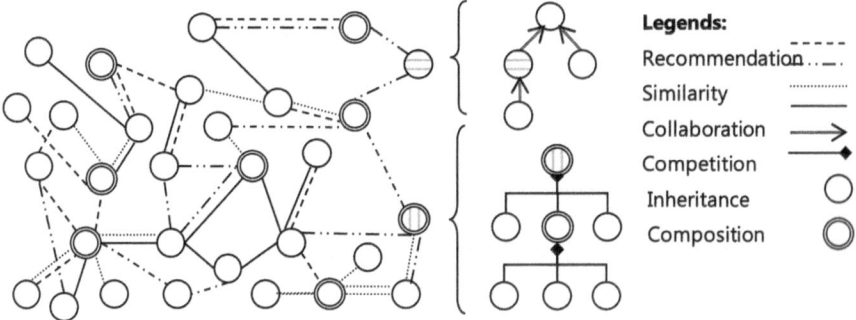

Fig. 4.2 A snapshot of relationships in digital ecosystem

In our research, we decide to give relationships between software components the same status as software components. We introduce the relationship pattern in Figure 4.1b to define various types of possible relations. Without loss of generalities, an entity refers to a software component, but it could be any element in the digital ecosystem (i.e., IT-enabled service). We define two categories of relationships between components to study interactions in digital ecosystems at the design time and their evolutions at the runtime. The first category comprises useful relationships to recursively build a "system of systems" by using the aggregation and the inheritance relationships, whereas the second category consists of relationships that imitate social-based networks of components, such as *recommendation, similarity, collaboration*, and *competition* relationships. To simulate a sustainable environment, the relationships patterns could enable social behaviors in digital ecosystems in which software components recommend alternative software to provide additional features. For examples, software components could recommend similar components to replace them in case of failures (see Fig. 4.2).

The **inheritance relationship** represents a generalization/specialization or parent/child relationship similar to the inheritance relationship in the object-oriented paradigm. The child component extends functionalities inherited from the base component. All child components may override base component functionalities.

The **composition relationship** refers to a whole/part relationship between a composite component and parts components. The composite component is a collection or container of subcomponents without specifying any execution order. The composition relationship enables reusability of parts that can be shared with other composite components.

The **recommendation relationship**: When building a composite component, each selected subcomponent may recommend peers to take part and enrich composition scenarios.

The **similarity relationship** indicates that some components can be substituted with peers providing similar functionalities if they fail at the execution time in order to maintain the ecosystem operational.

The **collaboration relationship** indicates that components with different functionalities can exchange information and build up composite components.

The **competition relationship** refers to components that have same functionalities and delivered by rival software providers. In this context, software components differ from each other by their QoS attributes, such as response time, availability, security, etc.

In digital service ecosystems, implicit or explicit, interactions and relationships between entities such as software components or IT-enabled services are useful in different ways: For example, we have used some relationships to capture implicit social links between Web services in order to enhance Web service discovery and define structural rules and dependency rules to respectively impose explicit constraints on control flows in composite Web services. Considering relationships between software components at design time and runtime in distributed and social-based environments provides new perspectives to enhance computational infrastructures, supporting IT-enabled services in digital service ecosystems.

4.3.2   The Evolution of Digital Service Ecosystems

Developing scalable digital ecosystem architectures to bundle IT-enabled services lead us to examine structures and behaviors of service systems as holistic and complex systems and understand relationships between service entities that cause service system evolution and, consequently, digital ecosystem evolution [48]. The evolution—or the capacity to be improved over time—partially depends on the evolution of digital entities and, especially, software components. Software components follow an evolutionary paradigm, which consists of the software life cycle such as software implementations, software usage, software upgrades, software archiving, or software abundance. The software disappears when it cannot evolve otherwise, as it is transformed by modifying its code and considering recent changes. By such, the evolution plays an important role in the sustainability of the digital ecosystem in response to changes. Understanding the evolution of digital ecosystems thus provides guidelines to scalable digital ecosystem architectures, such as what is the architecture style that ensure that the evolution of software components improves the digital ecosystem and does not lead to its disorder, how to capture the transition shift of digital ecosystem from one status to another during its evolution, can digital ecosystem evolution be described in terms of versions in a similar way to software versions, etc. For example, determining the version of a digital ecosystem poses a challenge, since all software components are distributed and run independently. Modifications in any software component may enable the transition of the digital ecosystem from current states to new states. Consequently, a new version can be assigned to the digital ecosystem to highlight this transition. To this end, we examine the circumstances, under which a new version is assigned to a digital ecosystem. Reasoning globally on software versions and evolutions explains their evolutions and is a key factor to understanding digital ecosystem versions.

1. Any system can be observed with respect to different and interrelated points of view. We study the digital service ecosystem with respect to three different points of view that we commonly refer to as being, doing, and becoming properties.
2. The *being property* is an ontological point of view, which considers the static structure of the system as a set of interrelated and organized entities. It describes the system's existence.
3. The *doing property* highlights the phenomenal or behavioral point of view, which considers system functionalities as sequences of actions.
4. The *becoming property* is a genetic point of view, which considers the system's evolution with respect to goal-driven rules and objectives. It controls all modifications related to being and doing properties and shows a self-awareness behavior regarding the system's evolution.

The evolution caused by changes due to the being property (i.e., informational flows, input and output data), the doing property (i.e., behaviors, processes, and procedures), or the becoming property (i.e., goals, policies, agreements) leads us to the definition of two evolution categories: *endogenous* and *exogenous evolution*.

4.3.3   The Service Bundling Based on Collaboration Processes

Most software architecture styles rely on interfaces (e.g., APIs, WSDL, RESTful APIs, ...) to interconnect and remotely invoke software components (e.g., EJB, modules, Web services, etc.). This increases software dependencies and causes interoperability problems when interfaces evolve. Middlewares reduce software dependencies on each other, but interfaces are still used to invoke software [238]. We propose a new architectural style to develop scalable digital ecosystem architectures to bundle IT-enabled services while preserving the expected service characteristics when software components evolve in IT-enabled services. We build the digital ecosystem architecture as an information-driven architecture with business artifacts as the basic construct to "connect" software components and, particularly, IT-enabled services. Our architecture is in line with our contributions to business artifact-driven service processes. The event-driven architecture is an example of an information-driven architecture, where events (i.e., changed states) are simple data that trigger tasks when they are exchanged [239]. Nevertheless, exchanged business artifacts are more than simple exchanged data and provide new capabilities, since they embody data, states, a life cycle, business rules, and tasks. Figure 4.3 depicts the business artifact-driven architecture to bundle IT-enabled service systems and, thus, connect their computational software infrastructures. By such, exposed service characteristics and their features facilitate IT-enabled service descriptions and discovery prior to bundling them, and business artifacts show how service goals are achieved without explicitly specifying software components that manipulate them (cf. interaction patterns). The connectivity of IT-enabled services with business artifacts ensures scalability and reduces complexity, since the business artifacts do not directly invoke software component interfaces.

The scalable digital ecosystem architecture is similar in spirit to the emergent architecture as recently proposed by Gartner [117]. The emergent architecture can be summarized as " [. . .] *'architect the lines, not the boxes', which means managing the connections between different parts of the business rather than the actual parts of the business themselves.*" This architecture advocates seven characteristic properties (e.g., non-eterministic and decentralized; autonomous, rule-bound, goal-oriented, and locally influenced actors; dynamic or adaptive systems; resource constraints) and provides insights to manage the growing variety and complexity in firms' information systems by modeling interactions between software components. Despite its characteristic properties and general approach, the emergent architecture has no design methods, architectural frameworks, or any implementation details.

To implement the digital ecosystem architecture to bundle IT-enabled services, we develop a generic architecture to implement business artifact-driven flows between IT-enabled services and a Software-as-a-Service (SaaS) architecture to provide service customers/partners a front-office interface to interact with the IT-enabled service based on the final service characteristics.

As illustrated in Fig. 4.4, the generic architecture relies on generic Web services whose roles are only limited to receiving and sending business artifacts. Since business artifacts are self-contained, each service providers define business rules to indicate how artifact attributes and states are updated and when artifact states are changed with respect to the artifact's life cycle. In such a way, service providers do not reveal their internal business activities, and thus they only guarantee that artifact will be processed based on current attribute values and states. However, partners can exchange business artifacts and sometimes wish to partially or totally hide their artifact contents. This typically leads us to develop the business artifact interaction patterns with dedicated flows. In this context, we differentiate between public and private artifacts to protect the privacy of each partner. Public artifacts are exchanged between partners, whereas private artifacts are internally managed by each partner. Public artifacts are considered reduced views of private artifacts. Consequently, partners should agree in advance on public artifact structures and their life cycles. The generic architecture includes the following elements:

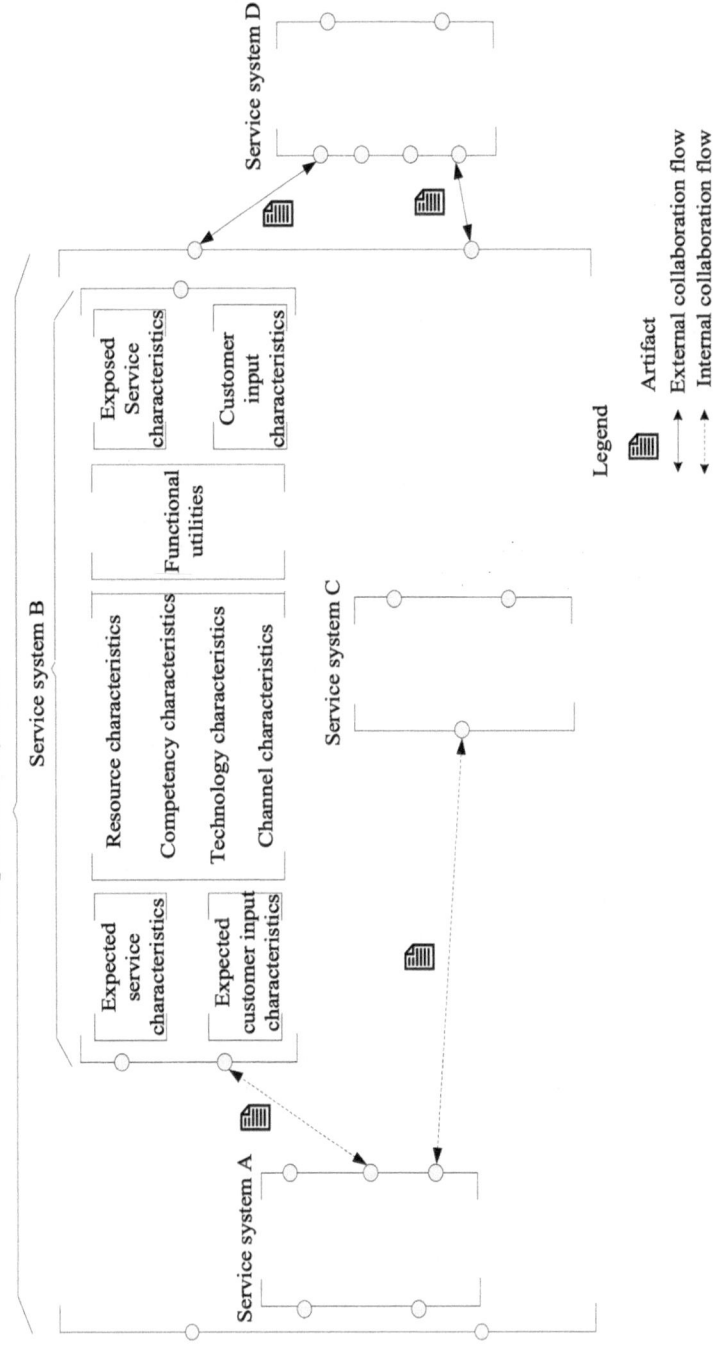

Fig. 4.3 Bundling IT-enabled service systems with business artifacts

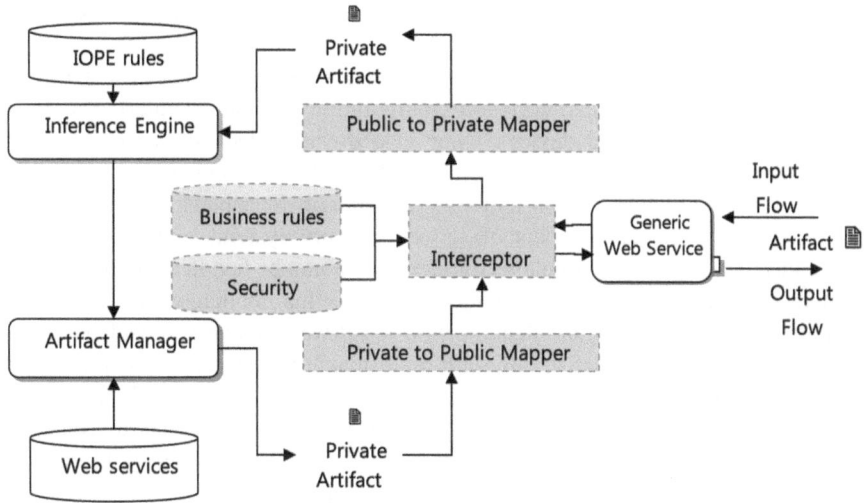

Fig. 4.4 Generic business artifact exchange architecture

- *Generic Web services*: are only used to facilitate the exchange of business artifacts encapsulated in input and output XML-based messages.
- *Business artifacts*: are modeled with any suitable information modeling approach such as entity-relationship diagrams. At the technical level, artifacts are implemented as XML schema.
- *Interceptor*: When the generic Web service receives a public business artifact, it invokes the interceptor to analyze artifact attribute values and states, check business constraints and requirements, and enable security credentials.
- *Artifact mappers*: Each partner has to define rules to specify how the mapping should occur between public and private artifacts.
- *IOPE databases*: Contain the ECA rules expressed as IOPE rules to invoke manual or automatic actions (i.e., Web services) that manipulate business artifacts.
- *Inference engine*: Transitions between artifact states are described in their artifact life cycles. Each transition can be realized by the execution of one or more actions. In this context, the inference engine reads the IOPE databases and deduces the rule that should manipulate the business artifact.
- *Artifact manager*: updates artifact and states according to business rules.

The interceptor and artifact mappers are optional components in the case that business artifact flows do not rely on private artifacts and there are no constraints to check when receiving artifacts.

4.3.3.1   The SaaS-Based Architecture for Service Front-Office Interfaces

Service customers, stakeholders, and users do not have same needs in terms of business rules, data models, software configurations, graphical interfaces, communication networks, and devices. They also do not select all service features and characteristics when they interact with IT-enabled services. In addition, they produce and share information and they collaborate to achieve common goals. In this context, the management of service delivery systems and service front-office interfaces become extremely complex and cannot be handled with traditional client/server or n-tier architectures. To overcome this challenge, we enable the digital service ecosystem architecture with a software delivery model based on Software-as-a-Service (SaaS) [240] with multi-tenant backends, in which multiple customers access shared data models and service characteristics based on licenses. Multi-tenancy refers to a single instance of a software component that runs on SaaS servers, serving multiple clients (i.e., tenants). With the multi-tenancy, IT-enabled service provides data and software virtualization capabilities.

The proposed SaaS-based delivery model illustrated in Fig. 4.5 enables the digital ecosystem architecture with front-office interfaces that allow service stakeholders to interact with IT-enabled services. We briefly introduce its modules as follows:

Tenancy partitioning module: customizes users' needs through data and software virtualization configurations. Each customer has a tenant with customized graphical user interfaces, private or public business artifacts, and shared data models. All tenants share the same SaaS instance that serves customers without any difference.

- *Feature bundle module*: allows a customer (i.e., tenant) to bundle service characteristics and features and select service utilities. This module runs a single software or process instance to serve all tenants and share data. Service stakeholders thus instantly benefit from new functionalities and releases when an instance is updated.
- *Contract management module*: manages service-level agreements (SLA) between customers and service providers. The SLA includes the expected service features and characteristics, key performance indicators, security, pricing plans, etc.
- *Service publishing module*: manages service characteristics and their features and service utilities in registries to facilitate their descriptions, discovery, and bundling.
- *Pricing catalog module*: manages monetizing models (e.g., pay-as-you-go, subscription, ...) for their offerings and managing revenues.
- *Monitoring module*: provides techniques, such as heartbeat monitoring and alert mechanisms, for problems caused by overloaded and/or crashed software components, network connections, or other devices. It also measures metrics to determine the status of availability, consistency and reliability, and SLA indicators.
- *Versionning management module*: measures the evolution of service characteristics from consumers and providers viewpoints as clarified in the next section.

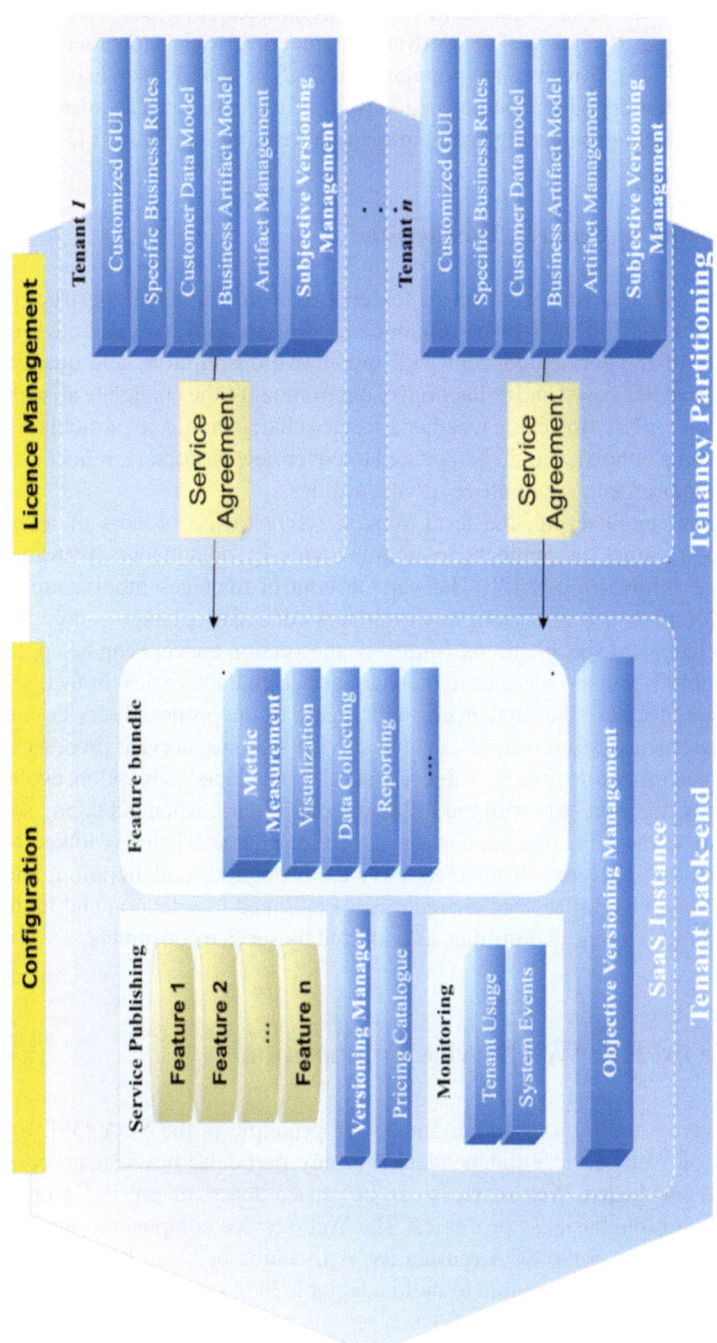

Fig. 4.5 SaaS-based IT-enabled service delivery interface

The implementation of the SaaS-based delivery interface model is not straight-forward in practice and requires advanced techniques. Fox and Patterson, for example, bring a diverse set of software engineering topics together for SaaS programming [241]. However, the popularity of SaaS is steadily increasing, because it supports customers with a single version of a software, reduces acquisition costs, and scales up without replacing costly infrastructure or adding IT staff [33].

4.3.3.2   Managing Service Characteristics

The service characteristic, as a set of features, is a fundamental construct in the characteristics view in the service reference model and aims at "concretizing" the intangibility of the service concept with qualitative descriptions and quantitative measures. In social-based and collaborative environments, the challenge arises when service providers have to decide whether a service characteristic, in particular within a multi-tenant's context, should be updated to cover new customers, reflect changes in service requirements, or improve service utilities.

In software engineering, the term version refers to the process of assigning either unique names or numbers to unique states of documents, software, and other digital information [242]. The version control manages modifications and code development, where a team of developers constantly change source files to implement technical specifications. Similarly, the version concept can be applied in the context of IT-enabled services to manage service characteristics. In fact, service consumers do not have the same needs with regard to the proposed services and its final characteristics. Moreover, IT-enabled service systems, service processes, and computational infrastructures are subject to service providers' perception, evolution, and revisions. We conclude with the following observation when managing service characteristics: the objective versioning of service characteristics is linked to the evolution of the delivered IT-enabled services throughout collaboration, and the subjective versioning of service characteristics is linked to selection and tuning of characteristics, features, and utilities as expected by service consumers.

4.4   The Ad Hoc Web Service Composition

The service reusability is an important design principle in the SOA [58]. It aims the design of Web services independently of any particular business process and technology, and hence, Web services can be reused across enterprise information systems in multiple business processes. The Web service composition mechanism plays a crucial role in the SOA reusability. A dynamic or an ad hoc composition mechanism has thus the potential to build adaptable SOAs.

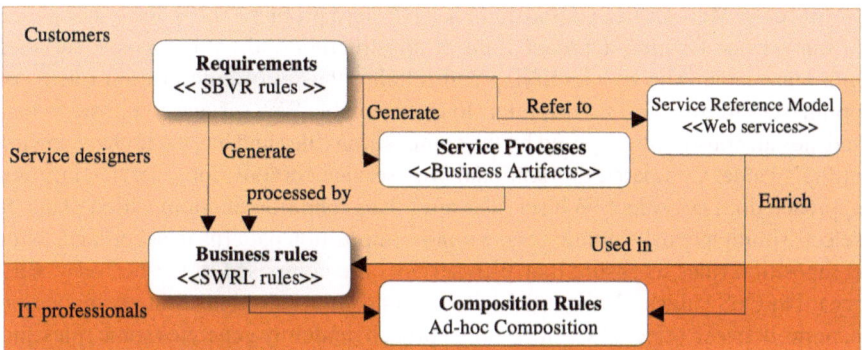

Fig. 4.6 From requirements to rules supporting ad hoc composition

In social-based, collaborative, and dynamic environments, IT-enabled services require an adaptable software infrastructures to ensure their adaptability to changes [22] and support business artifacts driven service processes [98]. SOA cannot provide adaptable infrastructures, unless an ad hoc Web service composition mechanism is developed to build adaptable SOAs. Only an ad hoc Web service composition mechanism can deal with changes, considering contextual information, service stakeholders' preferences, local and global constraints on functional and nonfunctional properties, implicit and explicit relationships between Web services, incomplete requirements, and dynamic recomposition of Web services without predefined composition plans.

To present the usage of the ad hoc Web service composition in service processes, we illustrate in Fig. 4.6 the generation of business artifacts and their business rules (SWRL rules) from service requirements expressed with SBVR (cf. requirement model). Business artifacts and collaboration patterns build up business processes (cf. collaboration model). Business rules (ECA) manipulate business artifacts by invoking appropriate actions (e.g., manual operations, select Web services or composite Web services, or build on-the-fly ad hoc composite Web services). In the case of composite Web services, changes in service requirements, collaboration scenarios, and the ICT-enabled environment may also trigger the composition of ad hoc composite web services without predefined composition plans.

The ad hoc service composition approach, called service farming, constructs composite services by simultaneously selecting atomic Web services and inferring composition patterns between selected services without a predefined composition plan. The service farming approach relies on a set of composition rules to capture respectively constraints from the ICT-enabled environment (e.g., relationships between Web services), service processes (e.g., business artifact rules), or the service system itself (e.g., set of available Web services, user preferences, quality of service...). Composition rules thus impact the ad hoc Web service composition process with a variety of rules, such as structural rules, dependency rules, local constraints rules, global constraints rules, and contextual rules.

Since the Web service composition is a NP-hard problem [220], the general idea of our service farming approach aims at constructing optimal composite services in a reasonable time by identifying and refining the set of composition rules. An optimal composite Web service refers to an acceptable Web service composition that satisfies all constraints imposed by the composition rules and has the highest service utility's value. Compared to traditional Web service compositions, service farming approach does not select Web services after the generation of composition plans. It selects simultaneously atomic services and composition patterns in accordance with composition rules to ensure that Web services are composed in the best way with regard to QoS values. Without the initial composition rules indicating composition patterns between selected services, our approach randomly generates some rules and enriches the set of composition rules at the end of each cycle. These rules will guide the following composition process in the upcoming cycles by using of the enriched set of composition rules. The composition process ends when an optimal composite service is found.

It is worth noting that the service farming approach includes abstract Web services and concrete/implemented Web services to express composition rules. An abstract Web service is a description for a generic concept or a capability, which can be realized or implemented by several concrete Web services.

4.4.1   The Rule-Driven Composition Model

The service farming approach is defined as a tuple: SF = <W, R, CM> where W is the set of Web services, R is the composition rule-based language (see Fig. 4.7), and CM is the composition model for a specific composition scenario expressed in terms of composition rules. The ad hoc composition algorithm applies rules in CM to Web services in W to construct composite Web services. We define the following Web services:

An **abstract service** a_i represents an action to be achieved, $a_i \in W_a$.
An **atomic service** t_j implements or realizes an abstract service, $t_j \in W_t$.
A **composite service** c_k consists of atomic services composed together, $c_k \in W_c$.

$W = W_a \cup W_t \cup W_c$, where W_a, W_t, and W_c are, respectively, the set of abstract services, atomic services, and composite services. s_i denotes a generic service from W_a, W_t, or W_c.

– The notation $t_j \sim a_i$ means that the atomic service t_j is one of the discovered services to realize the action represented by the abstract service a_i, $a_j \in W_a$, $t_j \in W_t$
– The notation $t_j \rhd c_k$ means that the atomic service t_j is one of the constituent atomic services of the composite service c_k. We assume that an atomic service is the constituent service of itself, denoted by $t_i \rhd t_j$, $a_i \in W_a$, $t_j \in W_t$, $c_k \in W_c$.

CompositionRules::= (CompositionRule, CompositionRules) | CompositionRules | ∅
CompositionRule ::= StructureRules | ConstraintRules | DependencyRules

StructureRules::= Service StructureOperator Service | Service StructureOperator StructureRule
StructureOperator:= ⊘sequence | ⓌParallel | ⊗Branch

DependencyRules ::= AService DependencyOperator AService
DependencyOperator::= ⊕Joint | +Optimize | ⊗Exclude

ConstrainRules::= LocalConstraint | GlobalConstraint | ContextualConstraint
LocalConstraint::= AService.QoSAttribute ConstraintOperator Value
GlobalConstraint::= CService.QoSAttribute ConstraintOperator Value
ContextualConstraint::= Service.QoSAttribute ConstraintOperator Parameter
ConstraintOperator::= < | ≤ | > | ≥ | = | #
Value::= Literal | Number Unit

Unit::= (Unit / Unit) | % | min | ...
Service::= AService | CService
AService::= Literal
CService::= Literal
QoSAttribute::= Literal
Parameter::= Literal
Literal::= [A-Za-z0-9#] | [A-Za-z]
Number::= [0-9#]*

Fig. 4.7 The composition rules model

The rule-driven language, R, defines three types of composition rules to represent the composition relations between different Web services, namely, structural rules, dependency rules, local constraints rules, global constraints rules, and contextual rules. Figure 4.7 illustrates composition rules in the context free grammar syntax.

1. **The structural rules:** provide information about the control flow between two Web services to be composed together. A structural rule is defined by using one of the following composition patterns:

   - **Sequence pattern**: sequential execution of s_i and s_j, denoted by $s_i \, s_j$
   - **Parallel pattern**: parallel execution of s_i and s_j, denoted by $s_i s_j$
   - **Selection pattern**: selective execution of s_i and s_j, denoted by $s_i s_j$
   - **Iteration pattern**: iterative execution of s_i, denoted by $!s_i$.

   In this context, each composite service c_i is regarded as a pair of atomic or composite services composed together following one of the composition patterns.

   A structural rule $r_i \in R$ consists of two services and their composition patterns, such as $r_i = <s_L, s_R, cp>$ where $s_L \in W$, $s_R \in W$ are two sub-services and cp ∈ {, , , !}

2. **The constraint rules:** set local constraints on individual atomic services and global constraints applied to all atomic services, participating in composite services. Three types of constraint rules are specified to filter services selected based on nonfunctional properties:

- **local constraint rules** express constraints on QoS attributes of individual Web services.
- **global constraint rules** set constraints on composite service QoS attributes.
- **Context constraint rules**: express constraints on abstract or atomic Web services based on contextual information defined by the collaboration context (e.g., preferences, location).

3. **The dependency rules:** capture explicit or implicit relationships among Web services to improve the selection of Web services prior to the composition process. We identify four types of dependency relationships:

   - **Composed rules**: s_i and s_j must be included in composite services, denoted by $s_i \oplus s_j$.
   - **Optimized rules**: When s_i and s_j are composed together, they should provide additional benefits or utility (e.g., lower price or lower costs), denoted by $s_i + s$.
   - **Excluded rules**: Either s_i or s_j is excluded and cannot be together in the same composite service, denoted by $s_i \otimes s_j$.

It is worth noting that structural rules and dependency rules express subtypes of social-based relationships between software components in the context of digital ecosystems, as defined in Sect. 4.3.3.1. The extensibility of composition rules with new rules is an important feature of our ad hoc rule-driven composition approach. New rules can be added to express additional requirements and constraints on the composition process. In this case, the service farming approach remains unchanged and proceeds the same way to find whether an optimal composite Web service can or cannot satisfy all composition rules.

4.4.2 The Service Farming Algorithm

The service farming approach is inspired by the data farming, which is an iterative process of developing, running, and analyzing large simulation models [243]. The objective of data farming is to generate and observe many possible outcomes and to obtain insights as to what factors drive the occurrence of each outcome [244]. It enhances the ability to discover trends and potential options in simulation results over extended ranges of input parameters and considers modeling and analyzing nonlinear phenomena with characteristics that cannot be precisely defined [245]. Similarly, the service farming algorithm consists of several cycles and generates a large number of composite Web services in each cycle. By following composition rules, composite Web services are obtained by randomly combining concrete and/or composite Web services from the previous cycle. In each cycle, composite Web services are ranked with respect to the highest service utility values. If the rank remains unchanged after several cycles, the top-ranked composite Web service is an optimal composition.

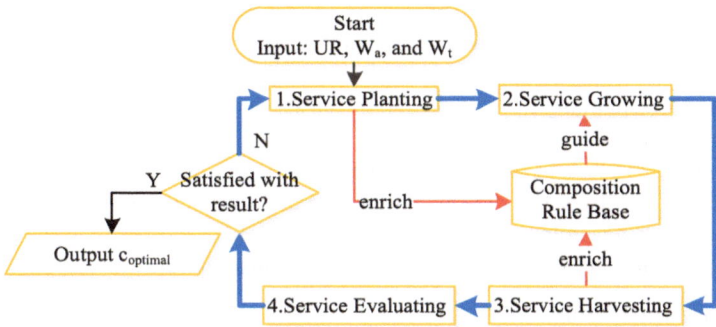

Fig. 4.8 The service farming stages

Driven by the composition rules, the service farming algorithm consists of four stages: service planting, service growing, service harvesting, and service evaluating, each of which is executed iteratively through several cycles to progressively construct composite services with atomic Web services and composite Web services from previous cycles. The flowchart of the service farming algorithm is shown in Fig. 4.8. We briefly describe hereafter the four stages with a pseudo-code syntax to illustrate the composition process.

1. **Service Planting Stage**

   The service planting stage focuses on generating composition rules issued initially from the composition model (CM) in the first cycle or based on enriched composition rules generated in previous cycles.

   Given the abstract service set W_a, the discovered atomic service set W_s, and the composite services W_c, this stage aims at:

   Step 1.   Specifying and generating the composition rules (e.g., structure rules, constraints rules, and dependency rules) based on the composition model CM.

   Step 2.   Initializing service seeds (W_{seed}) and service candidates ($W_{candidate}$) sets. The W_{seed} represents Web services that the algorithm uses to launch the composition process, whereas the $W_{candidate}$ refers to Web services that the algorithm randomly chooses to be composed with Web service in W_{seed}.

2. **Service Growing Stage**

   The Service Growing stage focuses on constructing composite services by incrementally applying composition rules to W_{seed}. In the case that rules are not available, two services are randomly composed following a parallel pattern or a sequential pattern as follows:

   - Step 1. For each service $s_i \in W_{seed}$, check for composed or optimal dependency rules, $r_{d.j}$ including s_i: if found, construct a composite service c_i matching $r_{d.j}$ else go to step 2.
   - Step 2. Pick $s_j \in W_{candidate}$, such as s_i and s_j are not in an excluded dependency rule.

- Step 3. Check structure rules $r_{s.k}$ indicating structure relationship between s_i and s_j: if found, construct a composite service c_i matching $r_{.j}$, or else go to step 4.
- Step 4. Randomly construct a composite service c_i with s_i and s_j in sequence or parallel.
- Step 5. If c_i satisfies functional requirements, then add it to composite service set W_c, or else select c_i as a service seed and start over from step 1.
- Step 6. When composition process is finished, R_c is applied to W_c to filter the composite services with respect to global constraint rules.

3. **Service Harvesting Stage**

   Based on W_c, the service harvesting stage firstly calculates for each service its utility value with the preference-based utility calculation formula (8). It applies the k-means cluster algorithm to divide W_c into several clusters based on composite services' QoS similarities and hence decomposes all composite services in the cluster with the maximum average utility value to update the global composition patterns table. The global composition patterns enrich $W_{candidate}$ and R_s in order to guide the service composition process in the next cycle. The harvesting stage includes the following:

- Step 1. Aggregate QoS of composite services.
- Step 2. Normalize QoS of composite services.
- Step 3. Calculate utility of composite services.
- Step 4. Cluster services in the composite service set W_c.
- Step 5. Calculate average utility of each cluster.
- Step 6. Find the cluster with maximum average utility W_{max}.
- Step 7. Decompose composite services from W_{max}.
- Step 8. Count appearance time.
- Step 9. Generate structure rules.

The QoS values of the service si are represented as a n-dimension vector such as: si.Q = < qi.1, qi.2, …, qi.n>, where qi.j represents the value of QoS attribute j of service si and n is the number of QoS attributes. To calculate QoS values of composite services, the following QoS aggregation equations based on the pessimistic model are used:

QoS	Sequence	And (concurrent)	Or (alternative)	Loop
Performance(P)	$\sum_{i=1}^{n} q_i^P$	$\sum_{i=1}^{n} q_i^P$	$\sum_{i=1}^{n} p_i q_i^P$	$k^* q_i^P$
Time(T)	$\sum_{i=1}^{n} q_i^T$	$\max\left\{q_{1,}^T, \cdots q_i^T\right\}$	$\sum_{i=1}^{n} p_i q_i^T$	$k^* q_i^T$
Availability(A)	$\prod_{i=1}^{n} q_i^A$	$\prod_{i=1}^{n} q_i^A$	$\sum_{i=1}^{n} p_i q_i^A$	$\left(q_i^A\right)^k$
Reliability(R)	$\frac{\sum_{i=1}^{n} q_i}{n}$	$\frac{\sum_{i=1}^{n} q_i^R}{n}$	$\sum_{i=1}^{n} p_i q_i^R$	q_i^R

In the following formula, the QoS normalization of composite services is achieved to transform different QoS attribute values into values between 0 and 1. The $q_{i.j}$ represents the value of QoS attribute j of the service s_i, and the $q'_{i.j}$

represents its normalized value, $s_i \in W_t \cup W_c$. In addition, $q_{best.j}$ and $q_{worst.j}$ are used to represent the best and worst values for QoS attribute j among all composite services.

$$q'_{i.j} = \left(\begin{array}{ll} \frac{|q_{\text{worst.j}} - q_{i.j}|}{|q_{\text{best.}} - q_{\text{worst.j}}|} & \text{if } q_{\text{best.j}} \neq q_{\text{worst}} \\ 1 & \text{else} \end{array} \right.$$

The normalized QoS of the service s_i is denoted by $s_i.Q' = <q'_{i.1}, q'_{i.2}, \ldots, q'_{i.n}>$, where $q'_{i.j} \in [0, 1]$, $s_i \in W_t \cup W_c$. The better the $q_{i.j}$ is, the more the value of $q'_{i.j}$ is close to 1.

The utility of service s_i is denoted as $s_i.u$, and it is calculated based on QoS preferences using:

$$s_i \cdot u = \frac{\sum_{j=1}^{n} \left((p_n - p_l \cdot q_{i.j} + 1) * q_{i.j} \right)}{\sum_{j=1}^{n} (p_n - p_l \cdot q_{i.j} + 1)}$$

In order to find the set, W_{max}, of composite services with the maximum average service utility value, we apply the k-means algorithm to cluster composite Web services in W_c based on their QoS vectors. The clustering algorithm relies on the euclidean distance to measure the QoS similarity between two composite services c_m and c_n.

After clustering, composite services in the same cluster have close utility values compared to services in other clusters. Regarding the cluster with the maximum average service utility value, the stage aims at finding structure patterns that make composite services possess higher utility values. It finally decomposes composite services into partial composition structures to enrich the structure rules by selecting the composition patterns with the highest appearance time.

4. **Service Evaluating Phase**

At the end of each cycle l_n, the composite service $c_{max}.l_n$ is found with the highest utility value $u_{max}.l_n$. Among all previous cycles, the optimal composite service $c_{optimal}$ is selected with the highest utility value $u_{optimal}$. If $u_{optimal}$ remains unchanged after a period t and cannot be improved anymore by the composition algorithm such as:

$$u_{max}l_n, u_{max}l_{n-1}, u_{max}l_{n-2}, \ldots, u_{max}l_{n-t} \leq u_{optimal}$$

The $c_{optimal}$ is thus the optimal composite Web service to be validated and executed.

4.5   From Service Systems to Digital Service Ecosystems

By mimicking the nature, which is supposed to be self-organized, self-managed, scalable, and able to provide complex solutions, recent works attempt to understand service systems as ecosystems by focusing on complex interactions between actors and entities immersed in technological, economic, political, and social environments. Ecology is the science concerning the interrelationship of living organisms and their environments and how organisms adapt to their surroundings [246]. The service, thus, is seen as some organisms cooperate with other organisms through intraspecific relationships to form communities and interact with their environment through interspecific relationships.

Based on similarities between services and ecological ecosystems, Mo and Wang study service systems from micro and macro perspectives by considering two domains (static structure domain and dynamic mechanism domain) and two levels (element level and system level). [196]. Golnam et al. propose a modeling framework called SEAM (Systemic Enterprise Architecture Method) to gain an understanding of how a service system maintains its identity and remains viable in its environment [96]. The viability of a system is a function of the balance between stability and adaptability. Saviano et al. [197] explores methodological links with service science and the viable system approach in order to interpret the emergent healthcare service systems instability. Barile and Polese examine similarities and differences of the fundamental principles of service-dominant logic and service science, and the viable systems model with a view to identify common features [29]. They argue that the viable systems approach provides insights into the design and management of service systems, especially with regard to systems governance. Tung and Yuan propose iDesign, a design framework of service systems by imitating symbiosis relationships in ecology to build communities that work together to survive [97]. The relationships between providers and consumers are driven by mutual adaptability associated with their behaviors when they engage in service exchange according to two conditions: continuity (mutualism, collaboration, commensalism) and mutual adaptability (one-sided, two-sided).

Designing and building service systems within a digital ecosystem framework is appealing. However, there is a lack of providing architectural styles, tools, and infrastructures to realize digital service ecosystems. In our research, we attempt to deal with this challenge and provide a bundling mechanism based on business artifacts. Establishing collaboration among IT-enabled services based on business artifacts provides a new approach to bundle services without dealing with technological interoperability. It also reduces the complexity of interactions among service stakeholders. Supporting IT-enabled service systems through the adoption of digital ecosystems offers new perspectives to explore a range of complex phenomena, many properties, and features found in the system thinking theory, such as the system of systems, complex systems, adaptive systems, etc.

4.6   Web Service Selection

Once Web services are discovered, selecting some of them based on nonfunctional properties is inevitably an essential activity before composing Web services. The Web service selection makes the Web service composition mechanism even more complicated and requires additional steps to identify the appropriate Web services based on their nonfunctional properties, such as quality of service (QoS) (e.g., availability, accessibility, response time, ...) and security (e.g., encryption, authentication, ...) [247]. Nonfunctional properties have substantial impacts on user expectations and can be used as discriminating factors among Web services providing similar functionalities. To incorporate QoS attributes in Web service selection, different approaches have been followed so far and can be classified into approaches that select individual Web services based on local constraints and without considering their composition context and approaches that select set of required Web services to be integrated into optimal Web service compositions based on global constraints to be satisfied by all selected Web services [248, 249].

Many approaches select Web services by comparing their nonfunctional properties to find out whether they are identical or similar [247, 250]. By doing so, the Web service selection process neglects nonfunctional property measurement scales, such as nominal, ordinal ratio, and interval scales, which require different comparison operators which exceed so far the equality operator. In addition, Web services, which are available in online UDDI registries, are not semantically annotated [65, 251, 252]. This is mainly because the WSDL is the standard language to syntactically describe Web services and does not provide semantic annotations with ontologies for describing nonfunctional properties. An overview of Web service selection approaches show divergence in solutions, making of them somehow difficult to be compared with each other. For example, Zhou et al. [68] propose a DAML-based QoS ontology for specifying various QoS properties and their metrics. Benbernou and Hacid [69] propose an ad hoc service description language to define complex constraints on QoS properties and incorporate them in a semantic Web service discovery process. Their resolution algorithm uses a rewriting system and the constraint propagation to evaluate queries over Web services. Maximilien and Singh [71] address dynamic Web service selections via an agent-based framework coupled with a QoS ontology. Nevertheless, they do not expose how they represent ontology-based QoS properties to Web service specifications. Diamadopoulou et al. [253] propose a Web service discovery model, which stores QoS information into either WSDL files or databases to be used by service selection intermediaries (i.e., brokers). Even though these approaches contribute to the Web service selection problem from different viewpoints and contexts, they remain customized or tailored solutions (e.g., insert QoS information in WSDL files or extend tModel fields or build ad hoc QoS models, etc.). By such, they are not conformed with existing Web service standards and recommendations, which cause incompatibility and interoperability problems. Despite extensions and improvements, UDDI registries do not provide mechanisms to publish and discover nonfunctional properties while preserving their structures and compatibility [254, 255]. They still grapple with inherent shortcomings such as the lack of semantics and contextual information [252]. In some situations, discovering Web services according to user preferences increases relevance of selected Web services that match user profiles. While most of Web service selection approaches focus on Web service nonfunctional properties in their matching algorithms, few approaches have dealt with user preferences to impact the matching process by assigning preference weights to the most preferred properties [72, 256]. These approaches hinge on semantic descriptions of user preferences as nonfunctional properties, making user preferences' management quite difficult.

Despite many interesting contributions, Web service selections still need standard-based solutions compatible with Web service recommendations and existing tools for wide acceptance by the software development community. The selection process should not oversimplify nonfunctional properties' scales and their corresponding operators and should provide a framework for managing nonfunctional properties from their definition, specification, and publication in UDDI registries to semantic selection with scale-based matching and negotiation capabilities.

In the following sections, we elaborate a novel Web service selection process to focus on nonfunctional properties (NFP). The process relies on a NFP ontology-based taxonomy, NFP measurement scales, and the semantic matching algorithm to consider user preferences selecting Web services and integrate them in the service farming algorithms.

4.6.1 Web Service Selection Based on Nonfunctional Properties

Web service selection requires a framework to manage Web service nonfunctional properties from specification to publication, selection, and negotiations, taking into account Web service standards, user preferences, and semantic matching with various nonfunctional property scales. In the following sections, we briefly present our contributions to Web service selection, which support the development of dynamic SOA with ad hoc Web service composition.

4.6.1.1 Web Service Nonfunctional Properties

Web service nonfunctional properties (NFP) vary from one domain to another, which leads to the lack of a global classification, facilitating their definition and manipulation. In response to this problem, we define a pivotal ontology-based taxonomy of NFPs to be referenced by Web services and used by the selection process. Each property belongs to a specific category, which can be either QoS-related or context-related see Fig. 4.9.

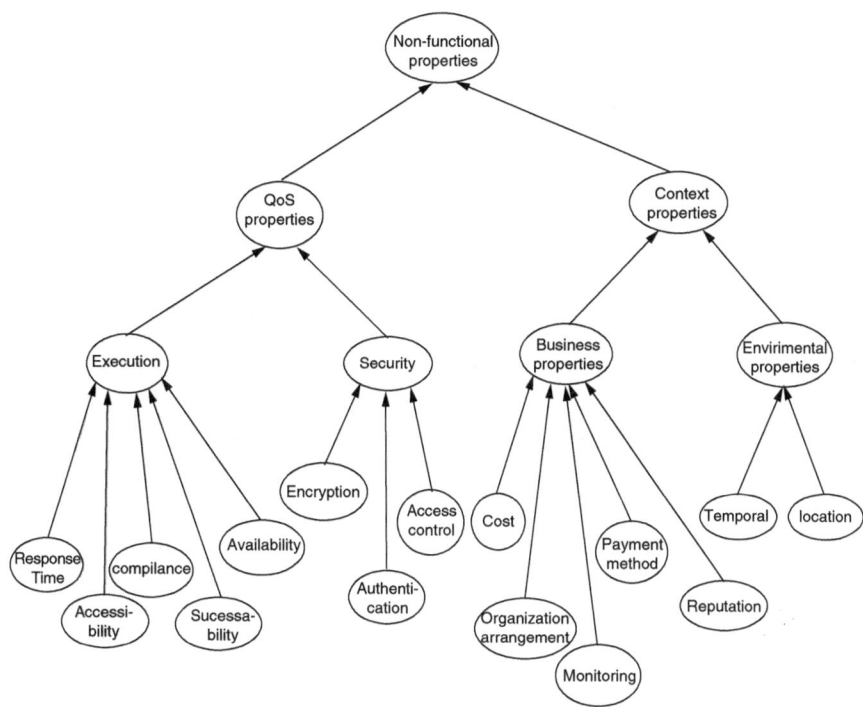

Fig. 4.9 The taxonomy of nonfunctional properties categories

The **QoS properties category** encompasses several quality parameters [257], which characterize behaviors of Web services based on their functionalities. It includes:

The **execution subcategory** includes the performance parameters, which characterize the interaction with the Web service, such as response time, availability, accessibility, successability, and compliance [258].

The **security subcategory** refers to the ability of a given Web service to provide suitable security mechanisms, such as authentication and access controls [259].

The **context properties category**: represents *"[. . .] a background knowledge useful in accomplishing some tasks. Task could be anything: solving a problem, reaching a conclusion, making a decision, answering a question, taking an action* [260]." Unlike QoS properties, context properties are relevant for differentiating Web services having same functional characteristics. It consists of two subcategories: environmental properties, which include location and temporal properties, and business properties, which include cost, reputation, payment method, and monitoring properties. From Web services point of view, the pivotal ontology-based taxonomy supports the Web service discovery process in distributed UDDI registries by providing a shared taxonomy. Yet another important issue in our selection framework is NFP measurement scales, which are often neglected by

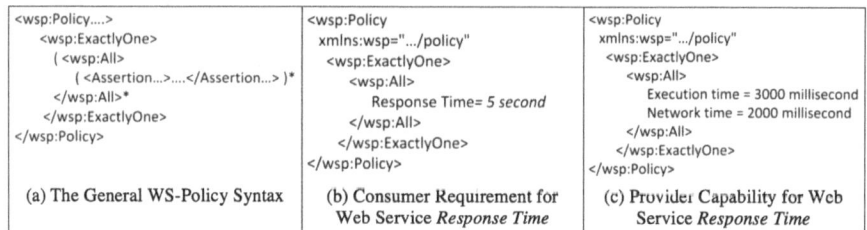

| (a) The General WS-Policy Syntax | (b) Consumer Requirement for Web Service *Response Time* | (c) Provider Capability for Web Service *Response Time* |

Fig. 4.10 The WS-policy assertion

Web service selection algorithms, which mainly match the similarity between Web service properties. We specify for each NFP a measurement scales as being nominal, ordinal, interval, or ratio scale. Measurement scales have direct implications on the type of analyses (e.g., matching or ranking) that should be performed on the NFP values.

- **Nominal scale:** refers to variables in which each value falls into one mutually exclusive and exhaustive category. For example, the "accessibility" has "private" or "public" value. They can only be compared through "=" and "≠" operators.
- **Ordinal scale:** indicates something about the rank-ordering of variable values in the sense that higher numbers represent higher values.
- **Interval scale :** When a nonfunctional property is measured on an interval scale, the distance between numbers or units on the scale is equal across the whole range of the scale.
- **Ratio scale:** Units are equal over the scale. It is also a meaningful zero point, which allows for the interpretation of ratio comparisons.

Considering measurement scales in matching Web properties entail specifications and implementations of various operators (e.g., interval comparison, set inclusion, etc.) to compare properties with respect to their scales. In our selection framework, measurement scales are specified with policy templates (Fig. 4.10) and used in the matching algorithm to compute distances to rank Web services (Fig. 4.12).

4.6.1.2   WS-Policy Specification to Model Nonfunctional Properties

As before mentioned, many Web service selection approaches attempt to specify NFPs with tailored solutions, UDDI extensions, ontology-based models, etc. As a result, these approaches fail to provide solutions compatible with Web service standards and recommendations and, consequently, decrease wide acceptance in a development context. In response to this challenge, we adopt the Web Services Policy (WS-Policy) [63] to define constructs to describe a broad range of service requirements and capabilities. The WS-Policy provides a grammar for representing Web services properties based on XML. A policy is defined as a collection of alternatives, which is, itself, defined as a collection of assertions (Fig. 4.10a). An assertion is used to represent a requirement, capability, or a behavior of a Web service [261]. Assertions specify characteristics, which are critical for selecting and using the Web services, for instance, contextual properties. An assertion can include an arbitrary number of child assertions and attributes. Nevertheless, WS-Policy presents some deficiencies in representing NFPs, because it only allows for a syntactic description of service properties. The example in Fig. 4.10b and c demonstrates that a syntactical matching between WS-policies of the response time nonfunctional property fails. In fact, a service consumer specifies the "response time" property as a requirement and a service provider expresses the "response time" property as two assertions: the "execution time" and "network time." A traditional matcher algorithm would perform a syntactic comparison or semantic matching of two ontology concepts and asserts that the two WS-Policies are different even though they are semantically compatible policies, if a rule inform that the response time can be deduced from multiplying the execution time and network time. This lack of semantic and reasoning on NFPs hampers the effectiveness of specifying and matching the compatibility between the WS-policies. To remedy this problem, we propose to extend WS-Policies for NFP with additional semantic information, by using common ontologies to allow semantic enrichment in the description of NFP policy assertions, and provide a matching algorithm enabled by transformation rules to assert that different syntactical assertions could be considered semantically compatible with respect to common ontologies.

To this end, we annotate WS-Policy assertions with the WS-Policy ontology to enable semantic and automatic parsing and reasoning over policies by means of transformation rules. In Fig. 4.11, the WS-Policy ontology shows that a policy belongs to a specific policy category, corresponds to a particular category (Policy-Category), and indicates the type of reasoning such as an assertion (MatchingType).

We distinguish two matching types (AssertionType); the "xsd" type denotes any data type supported by the XML schema such as string and float. In this case, the reasoning about the assertion similarity must consider the manipulation of such data type. The "ont" type indicates that a semantic reasoning must be applied to match this assertion. After defining WS-policy assertions, our matching algorithm assesses whether the reasoning based on transformation rules can be applied to assertions. In our previous example, policy assertions in Fig. 4.10b concerning the "Execution Time" and "Network Time" can be compared to the assertion "Response Time" in Fig. 4.10c by applying a transformation rule, Response_Time=Execution_Time+Network_Time, and setting its MatchingType to the "ont" value. The WS-Policy ontology provides the "Expression" concept to facilitate the matching of assertions in terms of "Parameter," "Unit," and "Operator." Another interesting feature of the WS-ontology is the FlexibilityMode concept that checks whether an assertion is negotiable (N) or non-negotiable (NN). By such, an NFP indicates that the Web service provider accepts to negotiate to modify its value. Finally, the "Scale" concept represents the measurement scale category defined in the previous paragraph. Finally, we assume that each Web service can have one or more NFP policies associated to each nonfunctional property.

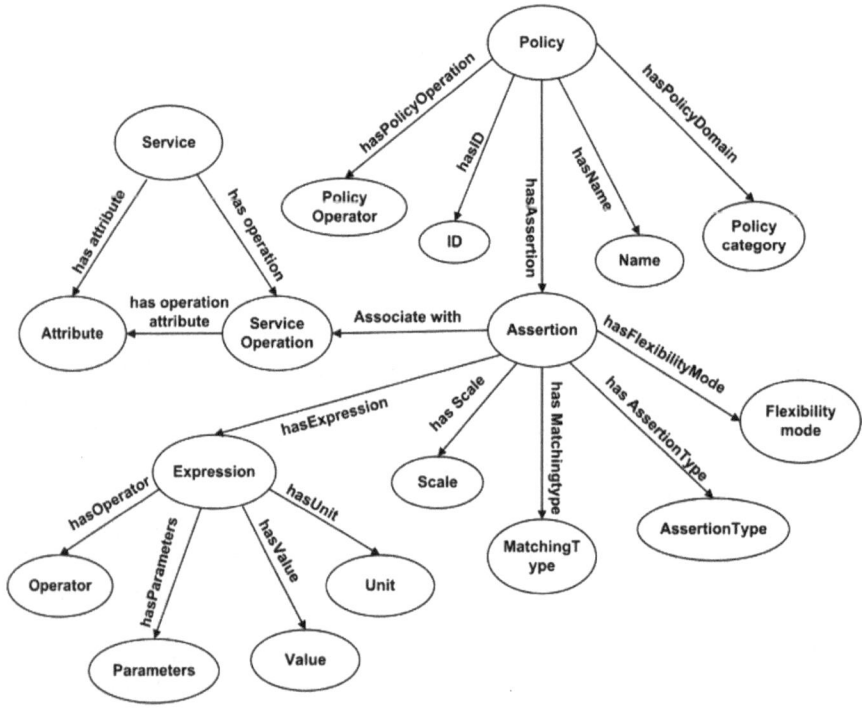

Fig. 4.11 WS-policy ontology

4.6.1.3 Publishing NFP WS-Policies in UDDI Registries

To disseminate NFP and respect Web service standards and recommendations, we employ UDDI registries to publish NFP WS-Policies through tmodels. A tModel is a generic data structure field that we use to register policies as a tuple: tModel=< ID_Policy, URL_Policy, ID_WS, Cg >, where Cg refers to the nonfunctional property category. In addition, we build communities of Web Services as containers to cluster together NFP policies in a particular category. A community is defined as a tuple: C= < ID_Community, Cg, memberList, scale, RuleSet>, where memberList attribute is defined as a tuple < ID_WS, ID_Policy > associating ID_WS to ID_Policy. The RuleSet= $\{R_i, \ldots R_n\}$ represents the set of transformation rules that extend the matching algorithm capabilities.

4.6.1.4   The Matching Algorithm

The core contribution in our framework is the algorithm that matches NFP policies of a given Web service (i.e., the source service) against all NFP policies of Web services available in the UDDI registry. The algorithm does not match NFP with operations associated to each measurement scales but also relies on transformation rules to infer new matching capabilities.

The algorithm is self-described in Fig. 4.12. The line (08) particularly shows that the matchNFPpolicy function matches the source NFP policy with all policy members of the community and relies on three phases: (i) the unification phase, (ii) the compatibility evaluation phase, and (iii) ranking and selection phase.

1. **The unification phase**: seeks to normalize policies, such as having same measurement units, and apply transformation rules to reason on assertions, such as transforming two policy assertions into an equivalent assertion to effectively compare them. The transformation rule for the example in Fig. 4.10 is expressed as following:

   If there exists a policy P, which has an alternative ALT, which has an Assertion A_1, which states that "ExecutionTime = X" and an Assertion A_2, which states that "NeworkTime = Y," **then** create a new Assertion A_3 which states that "ResponseTime = X + Y."

```
(01)      Function matchService (ID_WS)
              matchedWS= { };
(03)          response=1;
              WSPolicySet = getAllPolicy (Registry, ID_WS);
(05)          For P in WSPolicySet Do
                  PolicyCategory = getCategory(ID_Policy);
(07)              ID_Community = getCommunityID(PolicyCategory, communityList);
                  Invoke matchNFPpolicy(ID_Policy, ID_WS) Of ID_Community
(09)              candidateService = matchNFPpolicy(ID_Policy, ID_WS);
                  If response == 1 Then
(11)                  matchedService = candidateService;
                  Else
(13)                  matchedService = matchedWS ∩ candidateService;
                  response = response+1;
(15)          End DO
(16)      return (ID_WS, matchedWS)
```

Fig. 4.12 The matching algorithm

Table 4.1 The evaluation rule

Evaluation Rule template-name **NFP** **Category** NFP-category **Scale Type** scale-type **Policies** source-policy, member-policy **Action** evaluatePolicyCompatibility(source-policy, member-policy)	**EvaluationRule** template-ResponseTime **NFP** **Category** Response Time Scale Type Ratio **Policies** P_1, P_2 **Action** evaluatePolicyCompatibility(P_1, P_2) $if\ P_2 \le P_1 then\ return\ true\ if\ P_1 \le$ $P_2\ then\ P_1\ and\ P_2\ are\ negociable\ and$ $retunr\ Truelse\ return\ False$

2. **The evaluation phase**: Comparing two policies consists of comparing their
   assertions, respectively. The compatibility of two alternatives A and B denoted
   by ($A \equiv_{Alt} B$) is defined as follows: if and only if Capability Assertions (CA)
   of alternative A in the source policy match ($\perp_{Asser}$) the Requirement Assertions
   (RA) of alternative B in the request policy and the Requirement Assertions of
   alternative A match the Capability Assertions of alternative B:

   $(A \equiv_{Alt} B) \Leftrightarrow$

   $(\forall CA \in CA\,set(A), \exists RA \in RA\,set(B) S.T CA \perp_{Asser} RA) \wedge$

   $(\forall CA \in CA\,set(B), \exists RA \in RA\,set(B) S.T CA \perp_{Asser} RA)$

   CAset and RAset are, respectively, the set of capability and requirement assertions.

   In order to evaluate the WS-policy compatibility, the matchNFPpolicy function relies on evaluation rules to decide whether a member policy is compatible
   with a source policy. The syntax of an evaluation rule template is given below
   (Table 4.1):

   In the matching algorithm, the matchNFPpolicy function returns a list of
   candidate Web services, whose polices are compatible with the source policy
   or whose policies are not compatible, but they are declared negotiable by the
   service provider.

3. **Ranking and Selection Phase**

   The Web service selection process should identify the most appropriate candidate
   web service. The task is not simple, since each nonfunctional property has a
   measurement scale, which makes comparing all selected Web services together
   difficult to choose the most appropriate Web service. In response to this problem,
   we define a global distance to compute the dissimilarity between candidate
   services (i.e., candidate policies) and the source service (source policy). The
   service with the minimum distance is selected, because of closeness to the source
   Web service. We briefly introduce the hereafter ranking approach. Based on
   the set of Web services returned by the matching algorithm, we build the final
   selection matrix, $FSM_{ij} (1 \le i \le n, 1 \le j \le m)$, in which each
   row corresponds to a particular Web service while each column represents a

single NFP.

$$FSM = \begin{pmatrix} P_{1,1} & P_{1,2} & \cdots & P_{1,k} & \cdots & P_{1,n} \\ P_{2,1} & P_{2,2} & \cdots & P_{2,k} & \cdots & P_{2,n} \\ & & & & & \\ P_{n,1} & P_{n,2} & \cdots & P_{n,k} & \cdots & P_{n,m} \end{pmatrix}$$

The selection of services is based on the Minkowski distance between the services presented in FSM matrix and the service to be substituted:

$$d\,(S_i, S_S) = \sqrt{\sum_{k=1}^{m} g_k^2 \left(P_{i,k} - P_{s,k} \right)}$$

where s_i corresponds to the candidate service and s_s is a selected Web service returned by the Matching Algorithm. The g_k calculates the distances between the two NFP policies $P_{i,k}$ and $P_{s,k}$. $P_{i,k}$ is the policy k of the service i of the FSM matrix. $P_{s,k}$ is the policy k of the service s. It is worth noting that the g_k is a distance formula that depends on the nonfunctional property measurement scales.

4.6.1.5 Including User Preferences in Web Service Selection

We enhance the Web service selection process by including user preferences to leverage some nonfunctional properties, when they are critical from a user perspective or more important than other nonfunctional properties. Our contribution consists of applying the weighted sum method that parametrically changes the weights among objective functions to obtain the Pareto front. The initial work on the weighted sum method was done around 1963 by Zadeh [262]. For a set of objective functions $f_1, f_2, \ldots, f_n$. the weighted sum method takes each objective function and multiplies it by a weighting coefficient, w_i. The modified functions are then added together to obtain a single cost function, which can easily be solved using a single objective optimization method. The function is thus written as:

$$\sum_{i=1}^{n} w_i \cdot f_i \text{ where } 0 \le w_i \le 1 \text{ and } \sum_{i=1}^{n} w_i = 1$$

The trade-off surface may be generated by varying the weights $w_1, \ldots, w_n$. The main weakness of weighting technique is that it allows only linear relationships among objective functions and fails in the presence of non-convex search spaces. Consequently, it cannot generate all Pareto optimal solutions for non-convex trade-off surfaces. The weighting method is easy to implement. The weight itself reflects the relative importance (i.e., preference) among objective functions. Instead of specifying the real value for the nonfunctional properties, the user has to simply specify the importance of different NFPs, which makes this scheme easy to use.

Chapter 5
Services in the Era of Artificial Intelligence and Internet of Things

Keywords IoT-enabled services · AI-enabled services · Smart digital ecosystems · Smart service systems · Built-in data analytics · Self-adaptable services · Composable AI services · composable AI services · AI risks and services

5.1 Toward AI- and IoT-Enabled Services

According to the service-dominant logic, we discussed methods, models, and techniques to develop the digital service ecosystem, supporting IT-enabled services. The digital service ecosystem is based on challenges related to three prerequisites: service concept, service system, and service processes. We also applied the systemic thinking driven by a problem-solving approach [43] to explore the possibility of bringing coherence into various strands of services.

The research road map for a smart digital ecosystems will shift IT-enabled service systems to new levels and relies on devices, and AI analytics capabilities is illustrated in Fig. 5.1. The framework shows a multidisciplinary, integrative, and holistic research and development (R&D) approach. The framework comprises top-down and bottom-up R&D processes. The top-down R&D process starts with service engineering (business perspective) down to service engineering (technological perspective), while the bottom-up R&D process starts with cyber-physical system engineering (hardware perspective) to reach service engineering (technological perspective). In both processes, the goal is to find intermediary levels between service engineering and service computing and cyber-physical engineering and service computing, in which we can propose and develop solutions, combing principles, models, methods, and tools from IoT, AI, and cybersecurity disciplines. It is worth noting that cybersecurity and built-in data analytics are transversal challenges and should be holistically addressed at each level with solutions grounded in the hardware and software application to meet the business logic.

Given the global research framework, a multidisciplinary, integrative, and holistic R&D methodology is required to provide singular solutions relevant to smart

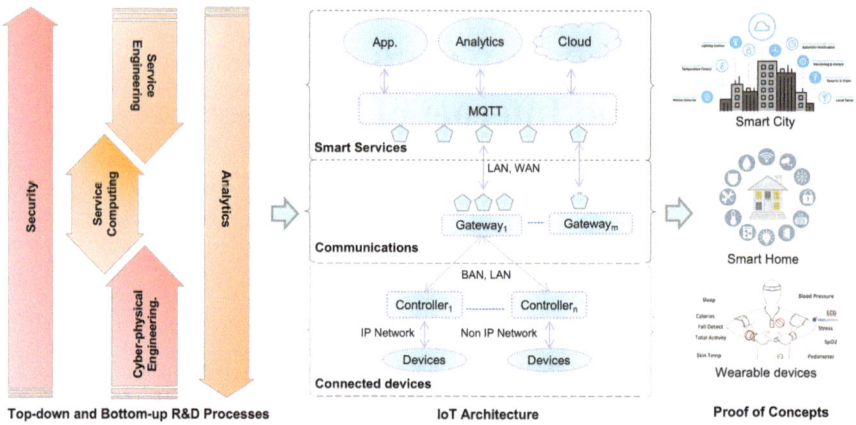

Fig. 5.1 A research framework for IoT-enabled services with data analytics capabilities

objects and smart services at a small scale (wearable medical devices), a medium scale (smart building), and a large scale (smart city). This chapter revolves around two main sections: Internet of Things and artificial intelligence. Each section comprises several directions and challenges from service systems perspectives. Cybersecurity is discussed from a transversal perspective across IoT and artificial intelligence.

5.2   Internet of Things and Services Systems

The premise behind the Internet of Things is to enable the interconnection of virtually any objects from the physical world to form networks, through which data is exchanged. The Internet of Things thus creates opportunities for more direct integration of the physical world into the Internet and results in efficiency and economic benefits in various fields, such as e-health, military, energy, urbanism, and businesses, just to mention a few. Connected things, also referred as smart objects, do not function independently; rather, they interact intelligently with each other without human interventions. Nowadays, reusability of connected devices in different scenarios and their abilities to behave smartly are lacking due to their software/hardware interoperability problems and the absence of built-in analytical capabilities to automatically adapt themselves to changes and/or provide new business insights at small and large scales. Since each smart object has limited computational and storage resources, privacy and security concerns surrounding the Internet of Things increase and require breakthrough solutions.

The emergence of connected devices (i.e., sensors, actuators, robots, digital artefacts, wearable devices, drones, and smartphones, etc.) offer tremendous opportunities to develop new services in everyday applications (i.e., e-health, remote assistance, tele-monitoring, telemetry, smart city, home automation, future factory). Even though networks of connected devices and big data analytics make possible to innovate in smart service design and delivery, we instantly realize that developing and deploying smart services for real-world business problems should address challenges related to IoT unconventional characteristics. Among these unconventional characteristics, we emphasize on the following:

- heterogeneity of hardware, software, communication protocols and exchanged datatype;
- device limited computational, storage and energy-power resources;
- large scale deployment in constrained and distributed environments;
- integration of social and cyber-physical requirements in the R&D process;
- need to continuously adapt connected devices to changes and resource consumptions;
- lack of devices' built-in analytic capabilities for decision-making supports and business insights;
- security and privacy challenges due these unconventional characteristics.

In the following sections, we identify four challenges to enable services with IoT capabilities and build interoperable, reusable, secure, and self-adaptable connected devices.

- Challenge 1: Interoperable and reusable connected devices
- Challenge 2: IoT big data management and built-in analytics

5.2.1   Self-Adaptable Connected Devices

Changes at runtime and inevitable evolution of user requirements make the management of connected devices a bottleneck for casual users (i.e., home automation, wearable devices) and even for professional operators (i.e., smart cities, smart grids, smart battle space). Scalability and wide usage of connected devices require self-adaptation mechanisms to drop the burden of manually adapting connected devices in response to changes. The control theory provides closed-loop controls for system automation. However, they are often carried out on subsystems to meet objectives (energy management, safety, automation, resilience, etc.) and thus remain limited to local optimums, which are sometimes contrary to a global optimum. As a result, self-adaptation in connected devices and smart services requires a paradigm shift from closed-loop controls to open-loop controls, making possible the emergence of optimal performance and coherent global behavior, regardless of whether devices are dynamically added, removed, or behaved abnormally. Self-adaptation in connected devices poses a scrupulous challenge and requires original contributions to handle different operational modes (synchronous vs. asynchronous coordination) and continuously monitor cyber and physical abnormal behaviors while coping with missing information in their surrounding environment. Device self-adaptation should consider deterministic behaviors driven by user requirements and business logics and nondeterministic behaviors with imperfect information to "learn" how connected devices should adapt their behaviors to unforeseen situations (exploiting edge machine learning algorithms). In many situations, IoT self-adaptation requires dynamic deployment of new connected devices or disposal of certain devices without interrupting the operations of service systems. The following challenges contribute toward the self-adaptation of devices in service systems:

1. A resilient architecture with proactive self-adaptive strategies driven by global and local QoS requirements (multi-objective optimizations) in response to contextual changes and dynamic environments. The self-adaptation relies on a declarative language for self-adaptation with deterministic behavior and a declarative language for self-adaptation with nondeterministic behavior and imperfect information based on machine learning algorithms under constrained resources.
2. A monitoring abnormal behaviors framework to detect malicious activities or malfunctioning through probes embedded in the hardware, communication sessions, and business logics.
3. A scalable M2M middleware for our component-based connected device model in order to support resilience, dynamic reconfiguration, fault tolerance, and de-authentication of devices in response to self-adaptation strategies.

5.2.2 IoT Big Data Management and Built-in Analytics

Given the rapid increase in the number of connected devices, their huge data volumes require big data infrastructures and specialized analytic tools. Nowadays, the management of big data in the IoT is only limited to load balancing, complex event processing, and efficient storage and processing clusters in the cloud or in centralized platforms. Such management oversimplifies the journey of data from sensors (sources) to storage and processing (sinks) and poses three major research problems:

Firstly, data analytic in the IoT occur in the cloud or in centralized systems on collected data (data in rest) or on streams of raw data (data in motion) generated by connected devices. In many situations, data analytic results and decisions should be sent back to devices to tune them or adjust them according to new business directives. The bidirectional data exchange is costly in terms of latency, power consumptions, and bandwidths and exposes transmitted data to security threats. In fact, not all data are meant to be sent to the cloud. In situations in which we require built-in data analytic capabilities under constrained resources to run on connected device or at the gateways' level, a new approach of embedding built-in analytic capabilities in connected device becomes an open research area of great interest for the machine learning community to run predictive algorithms with limited computational and storage resources (edge machine learning).

Secondly, big data management in the IoT relies on centralized platforms to directly or indirectly (via M2Ms) access each device to collect sensed data and, in few cases, change connected device settings or allow devices to request information from their peers (one-to-one access in push or pull modes). But, handling simultaneous and bulk accesses to many connected devices at large scale (i.e., smart city, smart grids, transport, airport automations, etc.) under conditions (i.e., location criteria, sensor types, value range, etc.) cannot scale up easily with individual accesses and cannot handle dynamics of devices.

In this context, there is a lack of (query) languages to massively manage devices from centralized IoT platforms or in peer-to-peer networks. Such (query) languages will support the built-in data analytic capability research topic and the need for a unified logical view of all connected devices. In addition, such language allows us to run locally built-in data analytic algorithms on devices and aggregating their results back to achieve real-time business analytics.

Thirdly, many third parties provide commercial platforms for IoT data analytics (AWS IoT services, Azure IoT suites, IBM IoT Watson IoT, ...) or open-source platforms (Kaa IoT platform, SiteWhere, ThingSpeak, DeviceHive, Thingsboard.io, DSA IoT platform, WSO2, ...). Most of these platforms feature components to set up devices and ingest, store, process, and integrate data from remote devices. However, they differ in many aspects in terms of their capabilities to conduct advanced data analytics, data visualization, and efficient management of connected devices. Most of them mainly focus on pulling data out of devices, storing data, performing analysis, and visualizing results on dashboards. However, analytic results are often insights and strategies that show new goals and possibilities to optimize global performance, increase benefits, or deduce risks in smart services and IoT-based systems. For now, it is unclear how these platforms can turn analytic results into commands to adjust device settings and business logics in order generate data to achieve new business goals issued from previous analytical results. In this context, feedback loops and cause-effect relationships on how data analytic results can affect the management of connected devices and their settings and, conversely, what data are required from connected devices to achieve desired business goals. The following challenges support IoT and big data management and built-in analytics in services systems.

1. Built-in data analytic algorithms to run on connected devices under constrained resources (edge machine learning) and integrate their capabilities in our reusable and interoperable connected device model.
2. A declarative (query) language to simultaneously manipulate connected devices at large scales based on filters and remotely invoke built-in data analytic algorithms embedded in devices before getting back their outcomes. This language should provide SQL-like statements to logically manipulate collections of devices without explicitly describing how to access them.
3. A big data analytics platform for the Internet of Things with capabilities to conduct analytics and turn analytical results into commands/actions that globally achieve desired goals.

5.3   Artificial Intelligence and Services Systems

Data analytics, machine learning, and artificial intelligence (AI) in general are increasingly becoming integral parts of many products and services that exhibit intelligent behaviors. The premise behind AI is to imitate human intelligence and/or automate cognitive tasks, such as visual perception, speech recognition, decision-making, translation between languages, planning, and reasoning. AI thus creates new opportunities from the integration of intelligent behaviors in service systems and results in efficiency and economic benefits in healthcare, finance, education, and hospitality, just to mention a few. Designing AI-centric service systems, therefore, has become and will be a norm in the future along with the proliferation of IoT and big data. These systems are likely to be distributed across machines and network boundaries with the potential to interact with external systems and adapt in response to self-learning capabilities (i.e., continual learning). Despite these advances and apart from Facebook, Apple, Amazon, Netflix, and Google, innovation in AI is still in its infancy and remains out of reach to individuals, scientists, entrepreneurs, investors, and public institutions when they attempt to train and test AI models at large scales. Challenges and barriers in AI innovation include (but not limited to) the following:

The Problem of Data Scarcity Data scarcity is one of the major bottlenecks for AI to reach production levels. Data collection is a difficult process, costly, and time-consuming. AI models require large amounts of data, and their performances heavily rely on data sizes and computational resources available for training and testing. Overfitting is a well-known phenomenon in machine learning when models overfit small training data and do not generalize well.

The Problem of Data Protection and Privacy A significant proportion of today's data (i.e., medical, financial, and personal data) on personal devices, within departments, and across institutions is not available for AI due to privacy requirements, ethical concerns, and legislation. Data privacy is a central concern in AI, especially when data contain personal, sensitive, and protected information. The ability to train AI models on sensitive data that remain with its owner (retention of sovereignty) may hold the key to unleash the potential of AI and cope with data scarcity and privacy.

Cybersecurity and AI Risks Cybersecurity and safety, if not addressed, expose AI-centric systems to vulnerabilities and threats and could pose risks to people's lives and to public safety. Not only should the correct functioning of AI-based systems be tested, but so should its resilience to adversarial attacks and how taking precautionary steps to mitigate risks. Risks could come from various sources, including deliberate attacks from adversaries, biases/unfairness in data and algorithms, events of unpredictable root cause, bugs in software, etc. Vulnerabilities could also be exploited to attack AI models during training, testing, and deployment by injecting backdoors, poisoning data, leaking private information, and altering their outputs. AI risk management and cybersecurity of AI perfectly align when there is a need to build trustworthy AI systems. These challenges and barriers lead to investigate research directions on trustworthy AI-enabled services and systems. The crucial question: How do we create an intelligence in services that can do what it was intended to do? In the following sections, we identify three challenges to enable trustworthy AI-enabled services.

- Challenge 1: Composable AI and security-by-design: How to develop AI capabilities by combining modular and reusable AI components (composable AI) while enabling security-by-design?
- Challenge 2: AI risk management and services: How to integrate AI risk management in AI-systems to identity, assess, and mitigate risks at model training and software development and detect risks at runtime?
- Challenge 3: Privacy-preserving and resilient federated learning: How to set up resilient and privacy-preserving federated learning in peer-to-peer environments to train on distributed data?

5.3.1 Composable AI Service Systems and Security-by-Design

Many of today's AI tasks are produced from machine learning (ML) pipelines, which consist of multiple sequential steps that are iteratively repeated to extract and preprocess data, train models, improve accuracy, and deploy or integrate them in large systems. For highly complex tasks, such as generating x-ray medical reports on behalf of doctors or processing multimodal presentation of medical records, ML pipelines are not trivial workflows and require specialized skills to produce hand-crafted solutions—only to discover that they cannot be scaled into production! Moreover, security and privacy requirements are often not considered or deemed difficult to apply to monolithic pipelines in collaborative projects. Recent works examined compositionality in ML[1] from reusable blocks to allow teams without advanced expertise in ML to build scalable pipelines. The block-based approach is a single step to easily build AI tasks, but the lack of security and privacy enforcement and the lack of integration with external services (e.g., Apps, business processes, devices) are open challenges.

In the context of AI-enabled services, these challenges leverage service computing to develop "composable AI" service systems by breaking down complex smart services, including AI/ML pipelines and AI related software, into loosely coupled services, which can be discovered, selected, and composed together while satisfying "security-by-design" patterns. To enable security-by-design, users should be able to specify their security preferences and requirements in services' profiles. Security requirements are then considered when discovering and selecting services

---

[1] Z. Liu et al. A Data-Centric Framework for Composable NLP Workflows, Proceeding of the 2020 Conference on Empirical Methods on Natural Language Processing. (EMNLP 2020 Demo).

and "correctly" deployed before composition. The development of theoretical foundations to specify AI components as reusable services and enable security-by-design requirements will enrich the body of knowledge of cybersecurity of AI and benefit from the Curry-Howard isomorphism, for example, to automate correct assembly (or composition) of AI components and generate composite components (i.e., smart services) through the standard relationship between the mathematical logic and ($\lambda - calculus$). The following challenges contribute toward security-driven composable AI service systems:

- Develop computational model and mathematical logic to uniformly represent AI services/components, and perform formal operations, such as validation, verification, automated composition, and safety testing.
- A modular and reusable Web service-based model and design method for ML pipelines and AI systems.
- AI-as-a-Service Cloud platform, integrating contributions from composable AI, AI risk management, and robust federated learning.
- Implement security mechanisms and protocols as services, specify user-security preferences and policies, and develop algorithms for matching security requirements and end-to-end deployment.

5.3.2 AI Risks in Service Systems

Many AI-centric service systems will be in tight interactions with humans and will offer unimagined opportunities in many areas (e.g., robotic surgery, driverless bus, fruit harvesting robots). Although undoubtedly beneficial, complex AI systems often exhibit emergent behaviors that manifest only through interactions with the world and with other systems in the environment. This implies that is impossible to predict with precision behaviors of intelligent service systems, unless we deploy and observe them. Unpredictability of AI is related to the impossibility results in computer science. For example, Rice's theorem states that "all non-trivial semantic properties of programs are undecidable" [263] and consequently, no method can "precisely and consistently predict what specific actions an intelligent system will take to achieve its objectives, even if we know terminal goals of the system." [264]. With unpredictability of AI in mind, the challenge is not only building capable intelligent service systems but also making them safe and secure. Unpredictability of AI increases AI threats due to unknown and known risks, such as biased data, adversarial attacks, software vulnerabilities, equipment failures, human machine errors, or environmental disruptions.

Even though achieving 100% safe AI is an impossibility, we can strive for safer AI services and identity methods to predict and mitigate potential risks, in particular cyber-threats (e.g., adversarial attacks, drift). Unfortunately, existing cybersecurity risk management frameworks (e.g., NIST CSF, Octave) are not applicable to AI due to AI intrinsic characteristics, such as cognitive and social behaviors, learning processes, fairness and biases, machine learning attacks, and vulnerabilities. A proactive approach to develop and integrate AI risk management framework in intelligent service systems should develop techniques to identify, assess risks and mitigate them in the design/training time, and monitor and treat risks in the runtime. These research topics could be extended to cover the whole spectrum of AI risks from a multidisciplinary perspective. For example, AI risks might interest researchers (social sciences, medicine), policymakers (legal, ethical, and political aspects), and computer scientists (learning theory, control theory, formal verification, AI testability, and expandability) at national, European, and international levels. AI risks is a controversial topic and has ethics concerns. On October 20,2021, the US National Institute of Standards and Technology (NIST) has organized its first workshop to help develop an AI risk management framework (800+ participants). In Canada, different institutions are conducting studies to answer questions regarding topics such as ethic and the risks and opportunities of AI innovation. The following challenges support AI risk management framework in services systems:

1. AI reference models: AI risks taxonomy, AI system model, cybersecurity/threat models
2. Methods to identify and assess risks, risk mitigation strategy, risks monitoring, and detection agent and National Catalog to publicly disclose ML Vulnerabilities & Exploits (similar to MRTE CVE)

5.3.3 Privacy-Preserving and Resilient Federated Learning

Privacy-preserving and data scarcity are the main motivation behind federated learning (FL). FL has emerged as a distributed approach to train centralized models with data across multiple edge devices and/or distributed servers. Local data is not exchanged and remain with its owner (data sovereignty). Although much effort has been put into FL from the ML perspective, user-centric security has been largely ignored when considering peer-to-peer and collaboration settings. In a peer-to-peer FL, end-users can become contributors (data owner) or initiators (model owner). In order to build privacy-preserving and resilient FL, my research investigates two interrelated research topics: user-centric security and cybersecurity of FL in trustless and distributed environments.

User-Centric Security Digital identity is the keystone upon which security services (i.e., authentication, authorization, secure exchanges) are built to ensure information security. In the context of FL settings (clients/server architecture), users rely on the server or third-party identity providers (PKI, OpenID, Shibboleth) to issue their identities (tokens, certificates), upon which they set authorization policies to control access to their local data and device resources. Hence, users have no control over their privacy (tracking user behaviors) and remain dependent on identity providers (i.e., a single point of failure). The integration of blockchains and AI and, in particular, the design of a blockchain-based user-centric security for decentralized service systems give stakeholders control to create and manage their digital identities, set up authentication, and define authorization policies to grant/revoke access to their data and devices. By incorporating user-centric security, FL becomes fully decentralized without trusting third-party identity providers. Recently, blockchain-based self-sovereign identity systems (uPort, Shocard, Sovrin) [265] have emerged but have not yet applied to FL. Recent efforts also investigate blockchains to build decentralized AI (SingularityNet, Microsoft SUM) [266] and improve cybersecurity of FL frameworks (BlockFedML, BlockFL architecture) [267]. Despite the synergy of leveraging blockchains to support AI and FL, these initiatives fail to simultaneously satisfy the triad of FL, adversarial defense, and user-centric security in peer-to-peer networks. Existing initiatives and solutions mainly rely on centralized architectures for FL, and users endure limited or no control over their data and devices in the realm of Internet of Things. This research topics aims at developing theoretical foundation and distributed algorithms for user-centric security using blockchain verifiable credentials for smart services.

Cybersecurity of Federated Learning Federated learning is not only subject to traditional security attacks (i.e., DDoS, eavesdropping, Sybil, and jamming attacks) and ML adversarial attacks (i.e., data poisoning, model inversion) but also new attacks to compromise the integrity of FL settings. Existing defense techniques, such as Byzantine-robust aggregation rules [268], focus on the aggregation process, where the server receives model updates from all clients and try to distinguish malicious updates from benign ones. These methods exploit statistical characteristics of model weights (trimmed mean, median, spectral anomaly detection) and often fail to detect backdoor attacks, because of non-IID distribution of data among different clients [269]. Pruning and fine-tuning techniques also proposed to mitigate backdoor attacks by eliminating neurons associated with extreme weights/inputs values. Studies show that pruning significantly degrade model performances on both clean data and back-doored instances [270]. Giving the significant growth of sophisticated attacks, it is difficult to decide which defense methods are more effective under different datasets, various data distributions, and attack types. Building a resilient artificial intelligence in distributed environments to train and test AI models urges research and business communities to consider cybersecurity of machine learning to aspire the development of trustworthy AI service systems.

References

1. CIA World Factbook, GDP Sector Composition: Field Listing—GDP Composition by Sector (2017). https://bit.ly/3Vcupf0. Accessed 7 Oct 2021
2. OECD, *Growth in Services Fostering Employment, Productivity and Innovation: Fostering Employment, Productivity and Innovation* (OECD Publishing, Berlin, 2005)
3. B.V. Looy, *Services Management: An Integrated Approach*, 2nd edn. (Financal Times Management, London, 2004)
4. L. Leydesdorff, *The Knowledge-Based Economy: Modeled, Measured, Simulated* (Universal Publishers, Irvine, 2006)
5. A. Breitenfellner, A. Hildebrandt, High employment with low productivity? The service sector as a determinant of economic development. Monetary Policy and the Economy, Oesterreichische Nationalbank (Austrian Central Bank) (1) (2006)
6. R. Rust, C. Miu, What academic research tells us about service. Commun. ACM **49**(7), 49–54 (2006)
7. J. Fitzsimmons, M. Fitzsimmons, *Service Management: Operations, Strategy, Information Technology*, 6th edn. (McGraw-Hill/Irwin, New York, 2007)
8. L.L. Berry, Services marketing is different. Business **30**(3), 24–29 (1980)
9. C. Schwengels, M. Opitz, Service systems engineering, in ed. by Deutsches Institut für Normung. Wege zur erfolgreichen Dienstleistungen (2005), pp. 22–45
10. Why We Need An Operations Paradigm for Services, Service-oriented computing: a research roadmap. Int. J. Coop. Inf. Syst. **17**(02), 223–255 (2008)
11. M.S. Daskin, *Service Science*, 1st edn. (Wiley, London, 2010)
12. S.E. Sampson, Why we need an operations paradigm for services, in *POMS/CSO Conference* (London business school, 2007)
13. G. Baxter, I. Sommerville, Socio-technical systems: from design methods to systems engineering. Interact. Comput. **23**(1), 4–17 (2011)
14. IfM and IBM, *Succeeding Through Service Innovation: A Service Perspective For Education, Research, Business And Government* (University of Cambridge Institute for Manufacturing, Cambridge, 2008)
15. N. Vrcelj, D. Trifunović, A. Jurvčiè, Distinctive characteristics of the service sector and its innovation challenges, in *Innovative Organization and Management* (2012), p. 113.
16. C.H. Lovelock, E. Gummesson, Whither services marketing? In search of a new paradigm and fresh perspectives. J. Serv. Res. **7**(1), 20–41 (2004)
17. J.M. Tien, D. Berg, A case for service systems engineering. J. Syst. Sci. Syst. Eng. **12**(1), 13–38 (2003)

© The Author(s), under exclusive license to Springer Nature Switzerland AG 2023   117
Y. Badr, *Smart Digital Service Ecosystems*, SpringerBriefs in Service Science,
https://doi.org/10.1007/978-3-031-27926-3

18. S.L. Vargo, R. Lusch, Service-dominant logic: continuing the evolution. J. Acad. Mark. Sci. **36**, 1–10 (2008)

19. S.E. Sampson, C.M. Froehle, Foundations and implications of a proposed unified services theory. Prod. Oper. Manag. **15**(2), 329–343 (2006)

20. B. Hefley, W. Murphy, R.F. Lusch, S.L. Vargo, The service-dominant mindset, in *Service Science, Management and Engineering Education for the 21st Century*, ed. by B. Hefley, W. Murphy, (Springer, New York, 2008), pp. 89–96

21. L. Schubert, K. Jeffery, B. Neidecker-Lutz, *A Roadmap for Advanced Cloud Technologies Under H2020* (Publications Office of the European Union, European Commission, Luxembourg, 2012), p. 35

22. P.P. Maglio, S. Srinivasan, J.T. Kreulen, J. Spohrer, Service systems, service scientists, SSME, and innovation. Commun. ACM **49**(7), 81–85 (2006)

23. H. Chesbrough, J. Spohrer, A research manifesto for services science. Commun. ACM **49**(7), 35–40 (2006)

24. J. Spohrer, S.L. Vargo, N. Caswell, P.P. Maglio, The service system is the basic abstraction of service science, in *Hawaii International Conference on System Sciences, Proceedings of the 41st Annual* (2008), pp. 104–104

25. R.F. Lusch, S.L. Vargo, G. Wessels, Toward a conceptual foundation for service science: contributions from service-dominant logic. IBM Syst. J. **47**(1), 5–14 (2008)

26. S. Alter, Service systems and service-dominant logic: partners or distant cousins? J. Relatsh. Mark. **9**(2), 98–115 (2010)

27. Y.-H. Tan, W. Hofman, J. Gordijn, J. Hulstijn, A framework for the design of service systems, in *Service Systems Implementation*, ed. by H. Demirkan, J.C. Spohrer, V. Krishna, (Springer, Boston, MA, 2011), pp. 51–74

28. R.G. Qiu, Computational thinking of service systems: dynamics and adaptiveness modeling. Serv. Sci. **1**(1), 42–55 (2009)

29. A Smarter Planet: Instrumented, Interconnected, Intelligen, F. Polese, Smart service systems and viable service systems applying systems theory to service science. Serv. Sci. **2**(1/2), 21–40 (2010)

30. S. Palmisano, *A Smarter Planet: Instrumented, Interconnected, Intelligent* (IBM, 2008). http://www.ibm.com/ibm/ideasfromibm/us/smartplanet/20081117/sjp_speech.shtml. Accessed 16 May 2012

31. D. Bhasin, *IT-Enabled Services* (Udaipur Chamber of Commerce, India; SPH Consultants, 2000)

32. S. Ovaska, J. Leino, *A Survey on Web 2.0.* (University of Tampere, Finland, 2008)

33. I. Churakova, R. Mikhramova, *Software As A Service: Study And Analysis Of Saas Business Model And Innovation Ecosystems* (Universiteit Gent, Ghent, 2010), p. 103

34. P. Dini et al., *The Digital Ecosystems Research Vision: 2010 and Beyond* (European Commission, Brussels, 2005)

35. S. Uesugi, IT-enabled services, in *IT Enabled Services*, ed. by S. Uesugi, (Springer, Vienna, 2013), pp. 1–17

36. P. Liu, P. Zhang, G. Nie, Business modeling for service ecosystems, in *Proceedings of the International Conference on Management of Emergent Digital EcoSystems*. New York (2010), pp. 102–106

37. M. Chau, G.L. Ball, J. Huang, J. Chen, J.L. Zhao, Global IT and IT-enabled services. Inf. Syst. Front. **13**(3), 301–304 (2010)

38. R.T. Rust, P.K. Kannan, *E-Service: New Directions in Theory and Practice* (M.E. Sharpe, Armonk, 2002)

39. B. Benatallah, O. Perrin, F. Rabhi, C. Godart, Web service computing: overview and directions, in *Handbook of Nature-Inspired and Innovative Computing*, ed. by A. Zomaya (Springer, New York, 2006), pp. 553–574

40. Y. Kim, K. Nam, Service systems and service innovation: toward the theory of service systems, in *Proceedings of the 15th Americas Conference on Information Systems, AMCIS 2009* (San Francisco, 2009)

41. L. Skyttner, *General Systems Theory. An Introduction* (Macmillan, London, 1996)
42. L.V. Bertalanffy, *General System Theory: Foundations, Development, Applications*, Revised (George Braziller, New York, 1969)
43. G. Bartlett, Systems thinking: a simple thinking technique for gaining systemic focus. Presented at the international conference on thinking, 2001
44. J.M. Wing, Computational thinking. Commun. ACM **49**(3), 33 (2006)
45. S.M. Goldstein, R. Johnston, J. Duffy, J. Rao, The service concept: the missing link in service design research?. J. Oper. Manag. **20**(2), 121–134 (2002)
46. P. Ralph, Y. Wand, A proposal for a formal definition of the design concept, in *Design Requirements Engineering: A Ten-Year Perspective* (Springer, Berlin, 2009), pp. 103–136
47. B. Edvardsson, J. Olsson, Key concepts for new service development. Serv. Ind. J. **16**(2), 140–164 (1996)
48. G. Briscoe, K. Keranen, G. Parry, Understanding complex service systems through different lenses: an overview. Eur. Manag. J. **30**(5), 418-426 (2012)
49. Z. Stanicek, M. Winkler, Service systems through the prism of conceptual modeling. Serv. Sci. **2**(1/2), 112–125 (2010)
50. C. Pinhanez, Services as customer-intensive systems. Des. Issues **25**(2), 3–13 (2009)
51. S. Alter, Service system fundamentals: work system, value chain, and life cycle. IBM Syst. J. **47**(1), 71–85 (2008)
52. J. Spohrer, P.P. Maglio, J. Bailey, D. Gruhl, Steps toward a science of service systems. Computer **40**(1), 71–77 (2007)
53. B. Monahan, D. Pym, R. Taylor, C. Tofts, M. Yearworth, *Grand Challenges for Systems and Services Sciences* (HP Labs, Palo Alto, 2006), p. 4
54. J.A. Buzacott, Service system structure. Int. J. Prod. Econ. **68**(1), 15–27(2000)
55. M. Bell, *Service-Oriented Modeling (SOA): Service Analysis, Design, and Architecture*, 1st edn. (Wiley, London, 2008)
56. E. Newcomer, *Understanding Web Services: XML, WSDL, SOAP, and UDDI*, 1st edn. (Addison-Wesley Professional, Boston, 2002)
57. P. Wohed, W. van der Aalst, M. Dumas, A. ter Hofstede, Analysis of web services composition languages: the case of BPEL4WS, in *Conceptual Modeling-ER* (2003), pp. 200–215
58. A. Dan, R.D. Johnson, T. Carrato, SOA service reuse by design, in *Proceedings of the 2nd international workshop on Systems development in SOA environments* (New York, 2008), pp. 25–28
59. B. Simon, Z. László, B. Goldschmidt, SOA interoperability, a case study, in *Proceedings of the IADIS International Conference, Informatics* (2008), pp. 131–138
60. Z. Gu, J. Li, R. Huang, Two extensions of service description to enhance the scalability of SOA, in *Proceedings of the 2nd International Conference on Scalable Information Systems* (ICST, Brussels, 2007), pp. 25:1–25:4
61. S. Tilley, D. Smith, H.A. Müller, Migrating to SOA: approaches, challenges, and lessons learned, in *The Conference of the Center for Advanced Studies on Collaborative Research*(Riverton, 2010), pp. 371–373
62. G. Kapitsaki, D.A. Kateros, I.E. Foukarakis, G.N. Prezerakos, D.I. Kaklamani, I.S. Venieris, Service composition: state of the art and future challenges, in *16th IST Mobile and Wireless Communications Summit* (2007), pp. 1–5
63. A. Tsalgatidou, T. Pilioura, An overview of standards and related technology in web services, Distrib. Parallel Databases **12**, 135–162 (2002)
64. L. Li, I. Horrocks, A software framework for matchmaking based on semantic web technology. Int. J. Electron. Commer. **8**(4), 39–60 (2004)
65. K. Verma, K. Sivashanmugam, A. Sheth, A. Patil, S. Oundhakar, J. Miller, METEOR-S WSDI: a scalable infrastructure of registries for semantic publication and discovery of web services. J. Inf. Technol. Manag. **6**(1), 17–39 (2005). Special Issue on Universal Global Integration
66. A. Sajjanhar, J. Hou, Y. Zhang, Algorithm for web services matching, in *Asia-Pacific Web Conference*, vol. 3007 (2004), pp. 665–670

67. D. Kourtesis, I. Paraskakis, in *Combining SAWSDL, OWL-DL and UDDI for Semantically Enhanced Web Service Discovery*. Lecture Notes in Computer Science, vol. 5021 (2008), pp. 614–628

68. C. Zhou, L. Chia, B. Lee, Service discovery and measurement based on DAML-QoS ontology. in *Special Interest Tracks and Posters of 14th World Wide Web Conference* (2005)

69. S. Benbernou, M.-S. Hacid, Resolution and constraint propagation for semantic web services discovery. Distrib. Parallel Databases **18**(1), 65–81 (2005)

70. X. Fan, C. Jiang, X. Fang, An efficient approach to web service selection, in *Web Information Systems and Mining* (Springer, Berlin, 2009), pp. 271–280

71. E.M. Maximilien, M.P. Singh, A framework and ontology for dynamic web services selection. IEEE Internet Comput. **8**(5), 84–93 (2004)

72. Z. Xu, P. Martin, W. Powley, F. Zulkernine, Reputation-enhanced QoS-based web services discovery, in *IEEE International Conference on Web Services* (2007), pp. 249–256

73. Q. Yu, M. Rege, A. Bouguettaya, B. Medjahed, M. Ouzzani, A two-phase framework for quality-aware Web service selection. Serv. Oriented Comput. Appl. **4**(2), 63–79 (2010)

74. C. Peltz, Web services orchestration and choreography. Computer **36**(10), 46–52 (2003)

75. R.K. Ko, S.S. Lee, E.W. Lee, Business Process Management (BPM) standards: a survey. Bus. Process. Manag. J. **15**(5), 744–791 (2009)

76. C. Ma, Q. Xu, J.W. Sanders, *A Survey of Business Process Execution Language (BPEL)* (United Nations University International—Institute for Software Technology, Macao, 2009)

77. Z. Baida, Software-Aided Service Bundling Intelligent Methods and Tools for Graphical Service Modeling. PhD Thesis. Amsterdam, Netherlands? Vrije Universiteit, vol. 301 (2006). ISBN: 10:90-810622-1-2

78. A. Arsanjani, S. Ghosh, A. Allam, T. Abdollah, S. Ganapathy, K. Holley, SOMA: a method for developing service-oriented solutions. IBM Syst. J. **47**(3), 377–396 (2008)

79. IBM Business Consulting Services, *Component Business Models Making Specialization Real* (IBM Institute for Business Value, 2005), p. 19

80. J. Cardoso, J.A. Miller, Internet-based self-services: from analysis and design to deployment, in *The 2012 IEEE International Conference on Services Economics (SE 2012)* (Hawaii, 2012)

81. T. Janner, C. Schroth, B. Schmid, Modelling service systems for collaborative innovation in the enterprise software industry—the St. Gallen Media reference model applied, in *2008 IEEE International Conference on Services Computing* (2008), pp. 145–152

82. J.A. Zachman, A framework for information systems architecture. IBM Syst. J. **26**(3), 276–292 (1987)

83. O. Zimmerman, P. Krogdahl, C. Gee, Elements of Service-Oriented Analysis and Design (2004). http://www-128.ibm.com/developerworks/library/ws-soad1/. Accessed 18 Mar 2013

84. W.D. Yu, C.H. Ong, A SOA based software engineering design approach in service engineering, in *Proceedings of the IEEE International Conference on e-Business Engineering*. Washington, DC (2009), pp. 409–416

85. M. Pistore, P. Traverso, M. Paolucci, M. Wagner, From software services to a future internet of services, in *Towards the Future Internet: A European Research Perspective* (ISO Press, 2009), pp. 183–192

86. H. Weigand, P. Johannesson, B. Andersson, J.A. Bergholtz, J.J. Arachchige, Closing the user-centric service coordination cycle, in *Proc of the 22nd International Conference on Advanced Information Systems Engineering (CAiSE'10) Forum* (2010)

87. Design Council, An Introduction to Service Design and a Selection of Service Design Tools (2012). www.innovateuk.org/_assets/pdf/design_methods_services.pdf. Accessed 18 Sep 2012

88. D. Vasiljevi, The service bundle design in a digital environment. Logist. J. Nicht-Referierte Veröffentlichungen (2008)

89. M.D. Eckersley, Designing human-centered services. Des. Manag. Rev. **19**(1), 59–65 (2008)

90. L. Lessard, E. Yu, A design theory and modeling technique for the design of knowledge-intensive business services, in *Proceedings of the 2012 iConference* (New York, 2012), pp. 510–512

91. J. Teixeira, L. Patrício, N. Nunes, L. Nóbrega, Customer experience modeling: designing interactions for service systems, in *Human-Computer Interaction—INTERACT 2011*, ed. by P. Campos, N. Graham, J. Jorge, N. Nunes, P. Palanque, M. Winckler, vol. 6949 (Springer, Berlin, 2011), pp. 136–143

92. R. Chase, D.E. Bowen, Service quality and the service delivery system, in *Service Quality: Multidisciplinary and Multinational Perspectives* (Lexington Books, Lexington, 1991), pp. 157–178

93. F. Ponsignon, P.A. Smart, R.S. Maull, Service delivery system design: characteristics and contingencies. Int. J. Oper. Prod. Manag. **31**(3), 324–349 (2011). https://doi.org/10.1108/01443571111111946

94. J. Sassoon, *Urbanisation des systèmes d'information* (Hermes Sciences Publications, England, 1998)

95. A. Erradi, S. Anand, N. Kulkarni, SOAF: an architectural framework for service definition and realization, in *Services Computing, 2006. SCC'06. IEEE International Conference on* (2006), pp. 151–158

96. A. Golnam, G. Regev, A. Wegmann, On viable service systems: developing a modeling framework for analysis of viability in service systems, in *Exploring Services Science* (Springer, Berlin, 2011), pp. 30–41

97. W.F. Tung, S.T. Yuan, iDesign: an intelligent design framework for service innovation, in *40th Annual Hawaii International Conference System Sciences*, vols. 3–6 (2007), pp. 64–74

98. W.M. Vanderaalst, D. Grunbauer, Case handling: a new paradigm for business process support. Data Knowl. Eng. **53**(2), 129–162 (2005)

99. S. Balin, V. Giard, A process oriented approach to the service concepts, in *International Conference on Service Systems and Service Management* (2006), pp. 785–790

100. C. Congram, M. Epelman, How to describe your service: an invitation to the structured analysis and design technique. Int. J. Serv. Ind. Manag. **6**(2), 6–23 (1995)

101. M.H. Boillot, G.M. Gleason, L.W. Horn, *Essentials of Flowcharting*, 5th edn. (Business and Educational Technologies, 1994)

102. R. Lu, S. Sadiq, A survey of comparative business process modeling approaches, in *Proceedings 10th International Conference on Business Information Systems (BIS), number 4439 in LNCS* (2007), pp. 82–94

103. S.E. Sampson, The unified service theory, in *Handbook of Service Science*, ed. by P.P. Maglio, C.A. Kieliszewski, J.C. Spohrer (Springer, New York, 2010), pp. 107–131

104. G.L. Shostack, Service positioning through structural change. J. Mark. **51**(1), 34–43(1987)

105. S. Kemsley, The changing nature of work: from structured to unstructured, from controlled to social, in *Business Process Management*, ed. by S. Rinderle-Ma, F. Toumani, K. Wolf (Springer, Berlin, 2011), pp. 2–2

106. J.L. Le Moigne, *La théorie du système général: théorie de la modélisation* (Presses Universitaires de France-PUF, Paris, 1994)

107. Ministry of Economy, Trade and Industry, Towards Innovation and Productivity Improvement in Service Industries. Commerce and Information Bureau Service Unit, Japan, (2007)

108. H. Cai, J.-Y. Chung, H. Su, *Relooking at Services Science and Services Innovation* (IEEE, Hong Kong, 2007), pp. 427–432

109. R. Pineda, A. Lopes, B. Tseng, O.H. Salcedo, Service systems engineering: emerging skills and tools. Procedia Comput. Sci. **8**(0), 420–427 (2012)

110. A. Finkelstein, J. Kramer, B. Nuseibeh, L. Finkelstein, M. Goedicke, Viewpoints: a framework for integrating multiple perspectives in system development. Int. J. Softw. Eng. Knowl. Eng. **2**(1), 31–58 (1992)

111. S. Holmlid, Interaction design and service design: expanding a comparison of design disciplines, in *2nd Nordic Design Research Design Conference*, vol. 11 (Stockholm, Sweden, 2009). Retrieved October, 2007

112. D. Matzen, A. Tan, M.M. Andreasen, Product/service-systems: proposal for models and terminology, in *Design for X* (Lehrstuhlfür Konstruktionstechnik, Erlangen, 2005), pp. 27–38

113. G.M. Weinberg, *An Introduction to General Systems Thinking*, 1st edn. (Wiley, London, 1975)
114. Defense Systems Management College, *Systems Engineering Fundamentals* (Defense Acquisition University Press, Fort Belvoir, VA, 2001)
115. M. Minsky, *Matter, Mind and Models, Rev. Version of the Essay in Semantic Information Processing* (MIT Press, Cambridge MA, 1995)
116. R.E. Giachetti, *Design of Enterprise Systems. Theory, Architecture, and Methods* (CRC Press, Boca Raton, 2010)
117. Gartner Inc., Gartner Identifies New Approach for Enterprise Architecture, Gartner Enterprise Architecture Summits, UK-2009. http://www.gartner.com/it/page.jsp?id=1124112. Accessed 2 Oct 2012
118. M. Cook, H.-P.P. Books, *Building Enterprise Information Architectures: Reengineering Information Systems*, 1st edn. (Prentice Hall, Englewood, Cliffs, 1996)
119. J. Schekkerman, *How to Survive in the Jungle of Enterprise Architecture Frameworks: Creating or Choosing an Enterprise Architecture Framework* (Trafford Publishing, Bloomington, 2003)
120. G. Scheithauer, Business service description methodology for service ecosystems, in *Proceedings of the 16th CAiSE-DC*, vol. 9 (2009), pp. 9–10
121. V. Bicer et al., Modeling Services using ISE framework: foundations and extensions, in *Modern Software Engineering Concepts and Practices: Advanced Approaches* (IGI Global, Pennsylvania, 2011), pp. 126–150
122. Y. Peng, Y. Badr, F. Biennier, Dynamic representation model for service systems, in *International Conference on Service Science (ICSS 2010)* (Hangzhou, 2010)
123. Y. Peng, Y. Badr, F. Biennier, A generic service system for knowledge-intensive service firms, in *The International ACM Conference on Management of Emergent Digital EcoSystems (MEDES)* (Lyon, 2009)
124. W. Li, Y. Badr, F. Biennier, Towards a capability model for web service composition, in *The IEEE 20th International Conference on Web Services (ICWS)*. (Santa Clara, 2013)
125. W. Li, Y. Badr, F. Biennier, Improving web service composition with user requirement transformation and capability model, in *The 21st International Conference on Cooperative Information Systems (CoopIS)* (2013)
126. Y. Peng, Y. Badr, F. Biennier, A pervasive environment for systemising innovative services in knowledge—intensive firms. Int. J. Electron. Bus. **9**(5), 429–453 (2011)
127. F. Biennier, R. Aubry, Y. Badr, A multi-dimensional service chain ecosystem model, in *International Conference on Advances in Production Management Systems (APMS)* (Bordeaux, 2009)
128. F. Gallouj, O. Weinstein, Innovation in services. Res. Policy **26**(4–5), 537–556 (1997)
129. K.J. Lancaster, A new approach to consumer theory. J. Polit. Econ. **74**(2), 133–157 (1966)
130. E. Mullera, D. Doloreuxb, What we should know about knowledge-intensive business services. Technol. Soc. **31**(1), 64–72 (2009)
131. K. Bhattacharya, N.S. Caswell, S. Kumaran, A. Nigam, F.Y. Wu, Artifact-centered operational modeling: lessons from customer engagements. IBM Syst. J. **46**(4), 703–721 (2007)
132. C. Fritz, R. Hull, J. Su, Automatic construction of simple artifact-based business processes, in *Proceedings of the 12th International Conference on Database Theory*. New York (2009), pp. 225–238
133. Z. Baida, J. Gordijn, H. Akkermans, A. De Boelelaan, A.Z. Morch, H. Sæle, Ontology-based analysis of e-service bundles for networked enterprises, in *Proceedings of the 17th Bled eCommerce Conference* (2004)
134. C. Rolland, Modeling the requirements engineering process, in *3rd European-Japanese Seminar on Information Modelling and Knowledge Bases V: Principles and Formal Techniques* (1993), pp. 85–96
135. J.M. Carrillo de Gea, J. Nicolás, J.L. Fernández Alemán, A. Toval, C. Ebert, A. Vizcaíno, Requirements Engineering Tools: Capabilities, Survey and Assessment. Inf. Softw. Technol. **54**(10), 1142–1157 (2012)

136. U. Nikula, J. Sajaniemi, H. Kälviäinen, *A State-of-the-practice Survey on Requirements Engineering in Small-and Medium-sized Enterprises* (Lappeenranta University of Technology Lappeenranta, Finland, 2000)

137. P. Donzelli, A Goal-driven and agent-based requirements engineering framework. Requir. Eng. **9**(1), 16–39 (2004)

138. M. Cohn, *Succeeding with Agile: Software Development Using Scrum*, 1st edn. (Addison-Wesley Professional, Reading, MA, 2009)

139. K. Schwaber, *Agile Project Management with Scrum*, 1st edn. (Microsoft Press, Unterschleissheim, 2004)

140. E. Yu, Modelling strategic relationships for process reengineering, in *Social Modeling for Requirements Engineering*, vol. 11 (MIT Press, Cambridge, 2011), p. 2011

141. K. Pohl, *The Three Dimensions of Requirements Engineering* (Citeseer, 1992)

142. S. Frolund, J. Koistinen, *QML: A Language for Quality of Service Specification, Technical Report HPL-98-10*, vol. 63 (Hewlett-Packard, Palo Alto, 1998)

143. A. Agrawal, Semantics of business process vocabulary and process rules, in *Proceedings of the 4th India Software Engineering Conference* (New York, 2011), pp. 61–68

144. I. Horrocks et al., SWRL: A Semantic Web Rule Language Combining OWL and RuleML. W3C (2011). http://www.w3.org/Submission/SWRL/

145. Y. Nakamur, S. Hada, R. Neyama, Towards the integration of web services security on enterprise environments, in *Proceedings of the Symposium on Applications and the Internet (SAINT) Workshops* (2002), pp. 166–175

146. S. Meilin, Y. Guangxin, X. Yong, W. Shangguang, Workflow management systems: a survey, in *International Conference on Communication Technology*, vol. 2 (1998), p. 6

147. L.M. Camarihna-Matos, H. Afsarmanesh, Concept of collaboration, in *Encyclopedia of Networked and Virtual Organizations*, ed. by G. Putnik, M. Cruz-Cunha (Information Science Reference, Hershey, 2008), pp. 311–315

148. K.L. Myers, P.M. Berry, others, At the boundary of workflow and AI, in *Proceedings of the Workshop on Agent-Based Systems in the Business Context* (1999)

149. J. Wang, A. Kumar, A framework for document-driven workflow systems, in *Business Process Management*, ed. by W. van der Aalst, B. Benatallah, F. Casati, F. Curbera, vol. 3649 (Springer, Berlin, 2005), pp. 285–301

150. D. Muller, M. Reichert, J. Herbst, Flexibility of data-driven process structures, in *Proceedings of the international conference on Business Process Management Workshops* (Berlin, 2006), pp. 181–192

151. S. Abiteboul, O. Benjelloun, T. Milo, The active xml project: an overview. VLDB J. **17**(5), 1019–1040 (2008)

152. T. Huckvale, M. Ould, Process modeling who, what and how: role activity diagramming, in *Business Process Change: Concepts, Methods and Technologies* (Idea Group Publishing, Harrisburg, 1995)

153. M. Weske, Formal foundation and conceptual design of dynamic adaptations in a workflow management system. Presented at the 34th annual hawaii international conference on system sciences, 2001, p. 10

154. S. Rinderle, M. Reichert, P. Dadam, Correctness criteria for dynamic changes in workflow systems— a survey. Data Knowl. Eng. **50**(1), 9–34 (2004)

155. D. Cohn, R. Hull, Business artifacts: a data-centric approach to modeling business operations and processes. IEEE Data Eng. Bull. **32**(3), 3–9 (2009)

156. A. Nigam, N.S. Caswell, Business artifacts: an approach to operational specification. IBM Syst. J. **42**(3), 428–445 (2003)

157. I. Beeson, S. Green, Using a language action framework to extend organizational process modelling, in *Proceedings of the 8th Annual Conference on Uk Academy for Information System* (2003)

158. T. Dufresne, J. Martin, Process modeling for e-business, in *INFS 770 Methods for Information Systems Engineering: Knowledge Management and E-Business* (George Mason University, Virginia, 2003)

159. W.M. Van der Aalst, M. Weske, D. Grünbauer, Case handling: a new paradigm for business process support. Data Knowl. Eng. **53**(2), 129–162 (2005)
160. H. de Man, *Case Management: A Review of Modeling Approaches* (Cordys, Netherlands, 2009)
161. E.J. Rooze, *ECase Management: An International Study in Judicial Organisations* (Netherlands Council for the Judiciary, Netherlands, 2007)
162. T. Chao et al., Artifact-based transformation of IBM global financing, in *Business Process Management*, ed. by U. Dayal, J. Eder, J. Koehler, H.A. Reijers, vol. 5701 (Springer, Berlin, 2009), pp. 261–277
163. K. Bhattacharya, C. Gerede, R. Hull, R. Liu, J. Su, Towards formal analysis of artifact-centric business process models. Business **4714**(13), 288–304 (2007)
164. Y. Peng, Y. Badr, F. Biennier, Designing data-driven collaboration in service systems, in *The 4th International Conference on New Trends in Information Science and Service Science (NISS)*. Gyeongju (2010).
165. N.C. Narendra, Y. Badr, P. Thiran, Z. Maamar, Towards a unified approach for business process modeling using context-based artifacts and web services, in *IEEE International Conference on Services Computing (ICSC 2009)*. Bangalore (2009)
166. Y. Badr, N.C. Narendra, Z. Maamar, Business artifacts for e-business interoperability, in *Electronic Business Interoperability: Concepts, Opportunities, and Challenges*, ed by E. Kajan (IGI Publisher, 2010)
167. Z. Maamar, Y. Badr, N.C. Narendra, Business artifacts discovery and modeling, in *8th International Conference on Service Oriented Computing*. San Francisco (2010)
168. E. Gamma, R. Helm, R. Johnson, J. Vlissides, *Design Patterns: Elements of Reusable Object-Oriented Software*, 1st edn. (Addison-Wesley Professional, Reading, MA, 1994)
169. G. Hohpe, B. Woolf, *Enterprise Integration Patterns: Designing, Building, and Deploying Messaging Solutions*, 1st edn. (Addison-Wesley Professional, Reading, MA, 2003)
170. V. Räisänen, W. Kellerer, P. Hölttä, O. Karasti, S. Heikkinen, Service management evolution, in *IST Mobile and Wireless Communications Summit*. Dresden (2005)
171. E. Gottesdiener, Decide how to decide: a collaboration pattern. Softw. Dev. Mag. **9**(1), 65–70 (2001)
172. A. de Moor, Community memory activation with collaboration patterns, in *3rd International Community Informatics Research Network (CIRN) conference*. Prato (2006), pp. 1–22
173. NSF IIS-0916515, Data-Centric Business Processes: Specification and Static Analysis (2009). http://db.ucsd.edu/artifacts/
174. S. Kumaran, R. Liu, F.Y. Wu, On the duality of information-centric and activity-centric models of business processes, in *Advanced Information Systems Engineering*, ed. by Z. Bellahsène and M. Léonard, vol. 5074 (Springer, Berlin, 2008), pp. 32–47
175. K. Bhattacharya, R. Hull, J. Su et al., A data-centric design methodology for business processes, in *Handbook of Research on Business Process Modeling* (IGI Global, Pennsylvania, 2009), pp. 503–531
176. B. Bagheri Hariri, D. Calvanese, G. Giacomo, R. Masellis, P. Felli, Foundations of relational artifacts verification, in *Business Process Management*, ed. by S. Rinderle-Ma, F. Toumani, K. Wolf, vol. 6896 (Springer, Berlin, 2011), pp. 379–395
177. D.B. West, *Introduction to Graph Theory*, 2nd edn. (Pearson, London, 2000)
178. J.F. Sowa, J.A. Zachman, Extending and formalizing the framework for information systems architecture. IBM Syst. J. **31**(3), 590–616 (1992)
179. H. Kautz, B. Selman, M. Coen, Bottom-up design of software agents. Commun. ACM **37**(7), 143–147 (1994)
180. P. Sabatier, Top-down and bottom-up approaches to implementation research: a critical analysis and suggested synthesis. J. Publ. Policy **6**(1), 21–48 (1986)
181. L. Silverston, W.H. Inmon, K. Graziano, *The Data Model Resource Book: A Library of Logical Data Models and Data Warehouse Designs*, 1st edn. (Wiley, London, 1997)
182. J. Küster, K. Ryndina, H. Gall, *Generation of Business Process Models for Object Life Cycle Compliance*, vol. 4714 (Springer, Berlin, 2007), pp. 165–181

183. A. Deutsch, R. Hull, F. Patrizi, V. Vianu, Automatic verification of data-centric business processes, in *International Conference on Database Theory* (2009), pp. 252–267
184. J. Rao, X. Su, A survey of automated web service composition methods, in *Semantic Web Services and Web Process Composition* (2005), pp. 43–54
185. Y. Charif, N. Sabouret, An overview of semantic web services composition approaches. Electron. Notes Theor. Comput. Sci. **146**(1), 33–41 (2006)
186. S. Yulu, C. Xi, A survey on QoS-aware web service composition, in *Third International Conference on Multimedia Information Networking and Security* (2011), pp. 283–287
187. E. Pejman, Y. Rastegari, P.M. Esfahani, A. Salajegheh, Web service composition methods: a survey, in *Proceedings of the International MultiConference of Engineers and Computer Scientists*, vol. 1 (2012)
188. K. Fujii, T. Suda, Dynamic service composition using semantic information, in *Proceedings of the 2nd International Conference on Service Oriented Computing*. New York (2004), pp. 39–48
189. W.-C. Chang, C.-S. Wu, C. Chang, Optimizing dynamic Web service component composition by using evolutionary algorithms, in *Web Intelligence, 2005. Proceedings. The 2005 IEEE/WIC/ACM International Conference on* (2005), pp. 708–711
190. D. Ardagna, B. Pernici, Adaptive service composition in flexible processes. IEEE Trans. Softw. Eng. **33**(6), 369–384 (2007)
191. L. Zeng, A.H. Ngu, B. Benatallah, R. Podorozhny, H. Lei, Dynamic composition and optimization of web services. Distrib. Parallel Databases **24**(1), 45–72 (2008)
192. P. Xiaoming, F. Qiqing, H. Yahui, Z. Bingjian, A user requirements oriented dynamic web service composition framework, in *International Forum on Information Technology and Applications* (2009), pp. 173–177
193. Z. Yang, C. Shang, Q. Liu, C. Zhao, A dynamic web services composition algorithm based on the combination of ant colony algorithm and genetic algorithm. J. Comput. Inf. Syst. **6**(8), 2617–2622 (2010)
194. B. Batouche, Y. Naudet, F. Guinand, Semantic web services composition optimized by multi-objective evolutionary algorithms, in *Proceedings of the 5th International Conference on Internet and Web Applications and Services* (Washington, DC, 2010), pp. 180–185
195. M. Huaxin, Study on Qos-oriented dynamic web service composition, in *IEEE 3rd International Conference on Communication Software and Networks* (2011), pp. 241—244
196. T. Mo, X. Xu, Z. Wang, A service system theory frame based on ecosystem theory, in *International Conference on Wireless Communications, Networking and Mobile Computing* (2007), pp. 3184–3187
197. M. Saviano, C. Bassano, and M. Calabrese, A VSA-SS approach to healthcare service systems: the triple target of efficiency, effectiveness and sustainability. Serv. Sci. **2**(1–2), 41–61 (2010)
198. A. Razavi, S. Moschoyiannis, P. Krause, An open digital environment to support business ecosystems. Peer Peer Netw. Appl. **2**(4), 367–397 (2009)
199. J.G. Miller, *Living Systems* (University Press of Colorado, Niwot, 1995)
200. G. Briscoe, P. De Wilde, Digital ecosystems: evolving service-orientated architectures, in *Proceedings of the 1st International Conference on Bio Inspired Models of Network, Information and Computing Systems* (New York, 2006)
201. G. Briscoe, P. De Wilde, Computing of applied digital ecosystems, in *Proceedings of the International Conference on Management of Emergent Digital EcoSystems* (New York, 2009), pp. 5:28–5:35
202. Directorate General Information Society and Media, Technologies for Digital Ecosystems—Innovation Ecosystems Initiative (2008). http://www.digital-ecosystems.org/index.htm. Accessed 7 Apr 2013
203. H. Fu, Formal concept analysis for digital ecosystem, in *5th International Conference on Machine Learning and Applications* (2006), pp. 143–148
204. A. Corallo, G. Passiante, A. Prencipe, *The Digital Business Ecosystem* (Edward Elgar Publishing, Cheltenham, 2007)

205. J. Cardoso, M. Winkler, K. Voigt, A service description language for the internet of services, in *Proceedings of ISSS* (2009)
206. D. Miorandi, S. Sicari, F. De Pellegrini, I. Chlamtac, Survey internet of things: vision, applications and research challenges. Ad Hoc Netw. **10**(7), 1497–1516 (2012)
207. S. Dustdar, W. Schreiner, A survey on web services composition. Int. J. Web Grid Serv. **1**(1), 1–30 (2005)
208. K.S.M. Chan, J. Bishop, L. Baresi, *Survey and Comparison of Planning Techniques for Web Services Composition* (University of Pretoria, Pretoria, 2007)
209. J. Peer, *Web Service Composition as AI Planning—A Survey* (University of St. Gallen, Switzerland, 2005) Second revised version
210. A. Strunk, QoS-aware service composition: a Survey, in *IEEE 8th European Conference on Web Services* (2010), pp. 67–74
211. M. Carman, L. Serafini, P. Traverso, Web service composition as planning, in *ICAPS 2003 Workshop on Planning for Web Services* (2003), pp. 1636–1642
212. L. Zeng, B. Benatallah, M. Dumas, J. Kalagnanam, Q.Z. Sheng, Quality driven web services composition, in *Proceedings of the 12th International Conference on World Wide Web* (New York, 2003), pp. 411–421
213. B. Benatallah, M. Dumas, Q.Z. Sheng, A.H. Ngu, Declarative composition and peer-to-peer provisioning of dynamic web services, in *18th International Conference on Data Engineering* (2002), pp. 297–308
214. H. Sun, X. Wang, B. Zhou, P. Zou, Research and implementation of dynamic web services composition, in *Advanced Parallel Processing Technologies* (2003), pp. 457–466
215. Y. Yang, S. Tang, Y. Xu, W. Zhang, L. Fang, An approach to QoS-aware service selection in dynamic web service composition, in *Third International Conference on Networking and Services* (2007), p. 18
216. L. Liu, A. Liu, Y. Gao, Improved algorithm for dynamic web services composition, in *The 9th International Conference for Young Computer Scientists, 2008. ICYCS 2008* (2008), pp. 342–347
217. G. Grossmann, R. Thiagarajan, M. Schrefl, M. Stumptner, Conceptual modeling approaches for dynamic web service composition, in *The Evolution of Conceptual Modeling* (Springer, Berlin, 2011), pp. 180–204
218. M. Vukovic, P. Robinson, Adaptive, planning based, web service composition for context awareness, in *International Conference on Pervasive Computing* (2004)
219. Q. Jiang, H. Xi, B. Yin, n-Online adaptive optimization for event-driven dynamic service composition. Control Theory Appl. **28**(8), 1049–1055 (2011)
220. N. Ben Mabrouk, S. Beauche, E. Kuznetsova, N. Georgantas, V. Issarny, *QoS-Aware Service Composition in Dynamic Service Oriented Environments*. Middleware (Springer, Berlin, 2009), pp. 123–142
221. T. Madhusudan, N. Uttamsingh, A declarative approach to composing web services in dynamic environments. Decis. Support. Syst. **41**(2), 325–357 (2006)
222. K. Ren, X. Liu, J. Chen, N. Xiao, J. Song, W. Zhang, A QSQL-based efficient planning algorithm for fully-automated service composition in dynamic service environments, in *IEEE International Conference on Services Computing*, vol. 1 (2008), pp. 301–308
223. W. Li, Y. Badr, F. Biennier, Digital ecosystems: challenges and prospects, in *Proceedings of the International Conference on Management of Emergent Digital EcoSystems* (2012), pp. 117–122
224. Y. Badr, Y. Peng, F. Biennier, Digital ecosystems for business e-services in knowledge-intensive firms, in *Business System Management and Engineering*, ed. by C.A. Ardagna, E. Damiani, L.A. Maciaszek, M. Missikoff, M. Parkin, vol. 7350 (Springer, Berlin, 2012), pp. 16–31
225. Y. Badr, G. Caplat, Software versioning and evolution in digital ecosystems, in *IEEE International Conference on Digital EcoSystems and Technologies (DEST)* (Dubai, 2010)

226. Y. Badr, G. Caplat, Software as a service and versionology: towards innovative service differentiation, in *The International Conference on Advanced Information Networking and Applications (AINA)* (Perth, 2010)

227. W. Li, Y. Badr, F. Biennier, Service farming: an Ad-hoc and QoS-aware web service composition approach, in *The 28th ACM Symposium on Applied Computing (ACM SAC)* (Coimbra, 2013), pp. 750–756

228. W. Li, Y. Badr, F. Biennier, A capability model for natural-language based web service composition, in *The 25th International Conference on Software and Systems Engineering and their Applications (ICSSEA).* (Paris, 2013)

229. Z. Maamar, N. Faci, L.K. Wives, Y. Badr, P.B. Santos, J.P.M. de Oliveira, Using social networks for web services discovery. IEEE Internet Comput. **15**(4), 48–54 (2011) (SCI, Impact Factor: 3.108)

230. Z. Maamar, L.K. Wives, Y. Badr, S. Elnaffar, K. Boukadi, N. Faci, LinkedWS: a novel web services discovery model based on the metaphor of social networks. Simul. Model. Pract. Theory **19**(1), 121–132 (2011)

231. Z. Maamar, Y. Badr, N. Faci, Q.Z. Sheng, Realizing an ecosystem of social web services: concepts, issues, and existing initiatives, in *Web Services Handbook*, ed. by B. Athman, Michael, D. Florian (Springer, Berlin, 2012)

232. Z. Maamar, N. Faci, Y. Badr, L.K. Wives, P.B. dos Santos, D. Benslimane, J.P.M. de Oliveira, Towards a framework for weaving social networks' principles into web services discovery, in *The International Conference on Web Intelligence, Mining and Semantics (WIMS'11)* (Finland, 2011)

233. Z. Maamar, D. Benslimane, Y. Badr, A context-based and policy-driven method to design and develop composite web services in transforming e-business practices and applications: emerging technologies and concepts, in *Advances in E-Business Research Series* (IGI Publisher, 2009), pp 385–406

234. S. Chaari, Y. Badr, F. Biennier, C. BenAmar, J. Favrel, Framework for web service selection based on non-functional properties. Int. J. Web Serv. Pract. **3**(2), 94–109 (2008)

235. S. Chaari, Y. Badr, F. Biennier, Enhancing web service selection by QoS-based ontology and WS-policy, in *Twenty-Third ACM Symposium on Applied Computing (ACM SAC)*, Ceará (2008), pp. 2426–2431

236. Y. Badr, A. Abraham, F. Biennier, C. Grosan, Enhancing web service selection by user preferences of non-functional features, in *Fourth International Conference on Next Generation Web Services Practices (NWeSP 2008)* (IEEE Computer Society Press, 2008), pp. 60–65. ISBN: 978-0-7695-3455-8

237. G. Morel, H. Panetto, F. Mayer, J.-P. Auzelle, System of enterprise-systems integration issues: an engineering perspective, in *IFAC Conference on Cost Effective Automation in Networked Product Development and Manufacturing* (2007)

238. D. Chappell, *Enterprise Service Bus: Theory in Practice* (O'Reilly Media, Sebastopol, 2004)

239. Z. Laliwala, S. Chaudhary, Event-driven service-oriented architecture, in *2008 International Conference on Service Systems and Service Management* (2008), pp. 1–6

240. K.S. Candan, W.-S. Li, T. Phan, M. Zhou, Frontiers in information and software as services, in *Proceedings of the 2009 IEEE International Conference on Data Engineering* (Washington, DC, 2009), pp. 1761–1768

241. A. Fox, D. Patterson, *Engineering Long-Lasting Software: An Agile Approach Unsing SAAS and Cloud Computing* (Strawberry Canyon LLC, California, 2012)

242. S.H. Kan, *Metrics and Models in Software Quality Engineering*, 2nd edn. (Addison-Wesley Professional, Boston, 2002)

243. G.E. Horne, K.-P. Schwierz, Data farming around the world overview, in *Proceedings of the 40th Conference on Winter Simulation* (2008), pp. 1442–1447

244. C.S. Choo, E.C. Ng, D. Ang, C.L. Chua, Data farming in singapore: a brief history, in *Proceedings of the 40th Conference on Winter Simulation* (2008), pp. 1448–1455

245. G.E. Horne, T.E. Meyer, Data farming: discovering surprise, in *Proceedings of the 36th conference on Winter simulation* (2004), pp. 807–813

246. T. Parsons, *The System of Modern Societies* (Prentice-Hall, Englewood, 1971)
247. M. Khezrian, W.M.N. Wan Kadir, S. Ibrahim, K. Mohebbi, K. Munusamy, S.G.H. Tabatabaei, An evaluation of state-of-the-art approaches for web service selection, in *Proceedings of the 12th International Conference on Information Integration and Web-based Applications and Services* (New York, 2010), pp. 885–889
248. D. Ardagna, B. Pernici, Global and local Qos guarantee in web service selection, in *Business Process Management Workshops* (2006), pp. 32–46
249. E. Vinek, P.P. Beran, E. Schikuta, Classification and composition of QoS attributes in distributed, heterogeneous systems, in *2011 11th IEEE/ACM International Symposium on Cluster, Cloud and Grid Computing (CCGrid)* (2011), pp. 424–433
250. M. Klusch, B. Fries, K. Sycara, OWLS-MX: a hybrid semantic web service matchmaker for OWL-S services. Web Semant. Sci. Serv. Agents World Wide Web **7**(2), 121–133 (2009)
251. M. Paolucci, T. Kawamura, T.R. Payne, K. Sycara, Importing the semantic web in UDDI, in *Proceedings of E-Services and the Semantic Web Workshop* (2002), pp. 225–236
252. L. Chen, Y. Li, R. Chow, Enhancing web service registries with semantics and context information, in *Proceedings of the 2010 IEEE International Conference on Services Computing* (Washington, DC, 2010), pp. 641–644
253. V. Diamadopoulou, C. Makris, Y. Panagis, E. Sakkopoulos, Techniques to support web service selection and consumption with QoS characteristics. J. Netw. Comput. Appl. **31**(2), 108–130 (2008)
254. P. Nasirifard, Web services security overview and security proposal for UDDI framework, in *Security and Management* (2003), pp. 348–351
255. C. Sun, Y. Lin, B. Kemme, Comparison of UDDI registry replication strategies, in *IEEE International Conference on Web Services* (2004), pp. 218–225
256. K. Zheng and H. Xiong, Semantic web service discovery method based on user preference and Qos, in *2nd International Conference on Consumer Electronics, Communications and Networks* (2012), pp. 3502–3506
257. R.L. Glass, Defining quality intuitively. IEEE Softw. **15**(3), 103–104 (1998)
258. M. Ouzzani, Efficient Delivery of Web Services. Ouzzani, Mourad. Efficient delivery of web services. Dissertation. Virginia Polytechnic Institute and State University (2003)
259. K. Khoo, L. Zhou, Managing web services security. J. Inf. Technol. Manag. **14**(3–4) (2004)
260. D. Martin, Putting web services in context. Electron. Notes Theor. Comput. Sci. **146**(1), 3–16 (2006)
261. K. Verma, R. Akkiraju, R. Goodwin, Semantic matching of web service policies, in *Workshop on Semantic and Dynamic Web Process Workshop* (2005)
262. L. Zadeh, Optimality and non-scalar-valued performance criteria. IEEE Trans. Autom. Control **8**(1), 59–60 (1963)
263. R. Péter. H.G. Rice, Classes of recursively enumerable sets and their decision problems. Trans. Am. Math. Soc. **74**, 358–366 (1953). J. Symbol. Logic **19**(2), 121–122 (1954). https://doi.org/10.2307/2268870
264. R.V. Yampolskiy, Unpredictability of AI: on the impossibility of accurately predicting all actions of a smarter agent. J. Artif. Intell. Conscious. **07**(1), 109–18 (2020). https://doi.org/10.1142/S2705078520500034
265. Y. Liu, D. He, M.S. Obaidat, N. Kumar, M.K. Khan, K.K. Choo, Blockchain-based identity management systems: a review. J. Netw. Comput. Appl. **166**, 102731(2020). ISSN 1084–8045
266. A.M. French, J.P. Shim, K.R. Larsen, M. Risius, H. Jain, The 4th industrial revolution powered by the integration of AI, blockchain, and 5G. Commun. Assoc. Inf. Syst. **49**(6), 266–286 (2021). https://doi.org/10.17705/1CAIS.04910

267. S. Wang, BlockFedML: blockchained federated machine learning systems, in *2019 International Conference on Intelligent Computing, Automation and Systems (ICICAS)* (2019), pp 751–756

268. S.P. Karimireddy, L. He, M. Jaggi, Learning from history for byzantine robust optimization, in *Proceedings of the 38th International Conference on Machine Learning, Proceedings of Machine Learning Research*, vol. 139 (2021), pp. 5311–5319

269. A.N. Bhagoji, S. Chakraborty, P. Mittal, S. Calo, Analyzing federated learning through an adversarial lens, in *Proceedings of the 36th International Conference on Machine Learning, Proceedings of Machine Learning Research*, vol. 97 (2019), pp. 634–643

270. S. Lu, R. Li, W. Liu, X. Chen, Defense against backdoor attack in federated learning. Comput. Secur. **121**(C), 102819 (2022). https://doi.org/10.1016/j.cose.2022.102819